MASTERMINDS Riddle Math Series

Elementary Grades

MEASUREMENT, FRACTIONS, PROBABILITY, AND LOGICAL THINKING

Reproducible Skill Builders And Higher Order Thinking Activities Based On NCTM Standards

By Brenda Opie
and Douglas McAvinn

Incentive Publications, Inc.
Nashville, Tennessee

Illustrated by Douglas McAvinn
Cover illustration by Douglas McAvinn

ISBN 0-86530-611-7

Copyright ©2004 by Incentive Publications, Inc., Nashville, TN. All rights reserved. No part of this publication may be reproduced, stored in a retrieval system, or transmitted in any form or by any means (electronic, mechanical, photocopying, recording, or otherwise) without written permission from Incentive Publications, Inc., with the exception below.

Pages labeled with the statement **©2004 by Incentive Publications, Inc., Nashville, TN** are intended for reproduction. Permission is hereby granted to the purchaser of one copy of **MEASUREMENT, FRACTIONS, PROBABILITY, AND LOGICAL THINKING** to reproduce these pages in sufficient quantities for meeting the purchaser's own classroom needs.

1 2 3 4 5 6 7 8 9 10   07 06 05 04

PRINTED IN THE UNITED STATES OF AMERICA
www.incentivepublications.com

TABLE OF CONTENTS

TIME SKILL BUILDERS

IT'S ALL ABOUT TIME! 2
The Time Machine 3
YOU ARE THE DESIGNER! 4
Making Time Work for You! 5
It's Roller Coaster Time! 6
Miss Froggy's Race to Beat the Clock! 7
Planning My Own Schedule 8
It Happened in the Barnyard! 9
Mr. Cricket's Watch Shop 10
INSTRUCTIONS FOR MATHO 11
MATHO 12
Putting Them In the Right Order! 13
READING A TIME LINE 14
MY TIME LINE 15
Important Inventions in the Last 200 Years 16

MONEY SKILL BUILDERS

Miss Froggy's School Shopping List 17
LIZZY NEEDS YOUR HELP! 18
Let's Go Shopping! 19
What's the Best Buy? 20
MAKING AN ORGANIZED LIST 21
MONEY TALK 22
BANANA SPLIT HEAVEN! 23
What's in the Treasure Chest? 24
Shapes, Shapes, and More Shapes 25
GET ABOARD! 26
You've Won a Shopping Spree! 27
The Detective's Assistant 28
What nursery rhyme do ghosts like best (Little Boo Peep) 29
Why couldn't the flower ride her bike to school?
    (Its petals were broken) 30
What happened when 2 frogs went after the same fly?
    (They became tongue tied) 31
What did the out of work skunk say?
    (I don't have a scent and it stinks) 32

What's in a Dollar?..33
   MONEY PROBLEMS!..34

STANDARD AND METRIC MEASUREMENT SKILL BUILDERS

Measurement Helper..36
Where do rabbits go when they get married?
   (On their bunnymoon)..37
MEASUREMENT CLOWN 38
"Mary, Mary, How Does Your Garden Grow?" 39
Where do fish keep their money? (In a river bank)...40
Why does a banana wear suntan lotion? (So it doesn't peel) 41
Why is the basketball player such a messy eater?
   (She dribbles all over the place) 42
It's All About Centimeters!..43
Metric Art Gallery 44
Let's Dig Into Metric Measurement!...45
METRIC SCAVENGER HUNT 46
MAKING THE BEST CHOICES!...47
STAR SEARCH IN CENTIMETERS 48
METRIC OCEAN HABITAT 49-50
MY METRIC HABITAT..51
What's the Temperature?..52
Prehistoric Creatures 53
WHAT AM I?..54
The Highs and Lows of It!...55-56
It's All About the Weather!..57-58
The Adventures of *Flat Stanley* 59-60

FRACTIONS AND PROBABILITY SKILL BUILDERS 61

Where in the World Can You Find Fractions? 62
Fractions in the "Real World" 63
When you can't fall asleep, why is it best to think about golf?
   (It putts you to sleep)...64
PUT ON A HAPPY FACE 65
CLOWIN' AROUND WITH FOURTHS!..66
Let's Take A Ride! 67
HEARTS GALORE! 68
CRAZY QUILTS 69
What did the dolphin say when it bumped into the shark?
   (I didn't do it on porpoise) 70

PROBABILITY SKILL BUILDERS
MAYBE, MAYBE NOT! 71
HEADS OR TAILS? 72
What's in the Pinata? 73

LOGICAL REASONING THINKING ACTIVITIES AND BRAIN TEASERS 74
TRICKS FOR SOLVING WORD PROBLEMS! 75
Animal Statistics 76
Where does a bunny rabbit go when her coat needs grooming? (To a hare dresser) 77
What's the Question? 78
What's the Question? – No. 2 79
Birthdays of Famous Baseball Players 80
What kind of insect do you swallow to relieve a cold? (Decongest-ant) 81
Case 1 – You're the Detective: What kind of potatoes are high in fat content? (Couch potatoes) 82
IT'S ALL IN WHERE YOU LIVE 83
What noise does a nut make when it sneezes? (Cashew) 84
What do aliens use to tie up spacemen? (Astroknots) 85
What did they call Old Macdonald when he joined the army? (GI GI Joe) 86
The Bake Sale! 87
It's Showtime!! 88
THE MONEY CHALLENGE 89

ANSWER KEY 90-91

1

TIME AND MONEY SKILL BUILDERS

©2004 by Incentive Publications, Inc., Nashville, TN.

Connecting math to the real world NAME_____ 2

IT'S ALL ABOUT TIME!

Think of 5-7 ways your life might be different if there were no clocks.

1. _____
2. _____
3. _____
4. _____
5. _____
6. _____
7. _____

Now, think of many, varied, and unusual places you have seen clocks or other timepieces in the real world.

1. _____   5. _____
2. _____   6. _____
3. _____   7. _____
4. _____   8. _____

©2004 by Incentive Publications, Inc., Nashville, TN.

3 Making and using a clock

NAME_____

The Time Machine

Make your own time machine by using the patterns below. You can then practice telling time by moving the hands on your clock.

minute hand

hour hand

brad

©2004 by Incentive Publications, Inc., Nashville, TN.

Using creativity

NAME_____

YOU ARE THE DESIGNER!

There are many kinds of clocks: cuckoo clocks, Mickey Mouse clocks, clocks shaped like jets, Barbie clocks, and clocks that look like soccer balls. In the space below, you get to design your own clock. Try to think of many, varied, unusual designs and draw the one you think is the most unusual.

Write 2-3 sentences telling about your clock. _____

_____

_____

©2004 by Incentive Publications, Inc., Nashville, TN.

5 Correlating time with daily activities   NAME_____

Making Time Work for You!

Paste or draw a picture on the chart to show what you might be doing at each of the times given in the boxes below.

7:30 AM	3:30 PM
11:30 AM	6:00 PM
2:00 PM	9:00 PM

©2004 by Incentive Publications, Inc., Nashville, TN.

Telling time

NAME _____

It's Roller Coaster Time!

After you have read each sentence, draw hands on the clock to show the correct time.

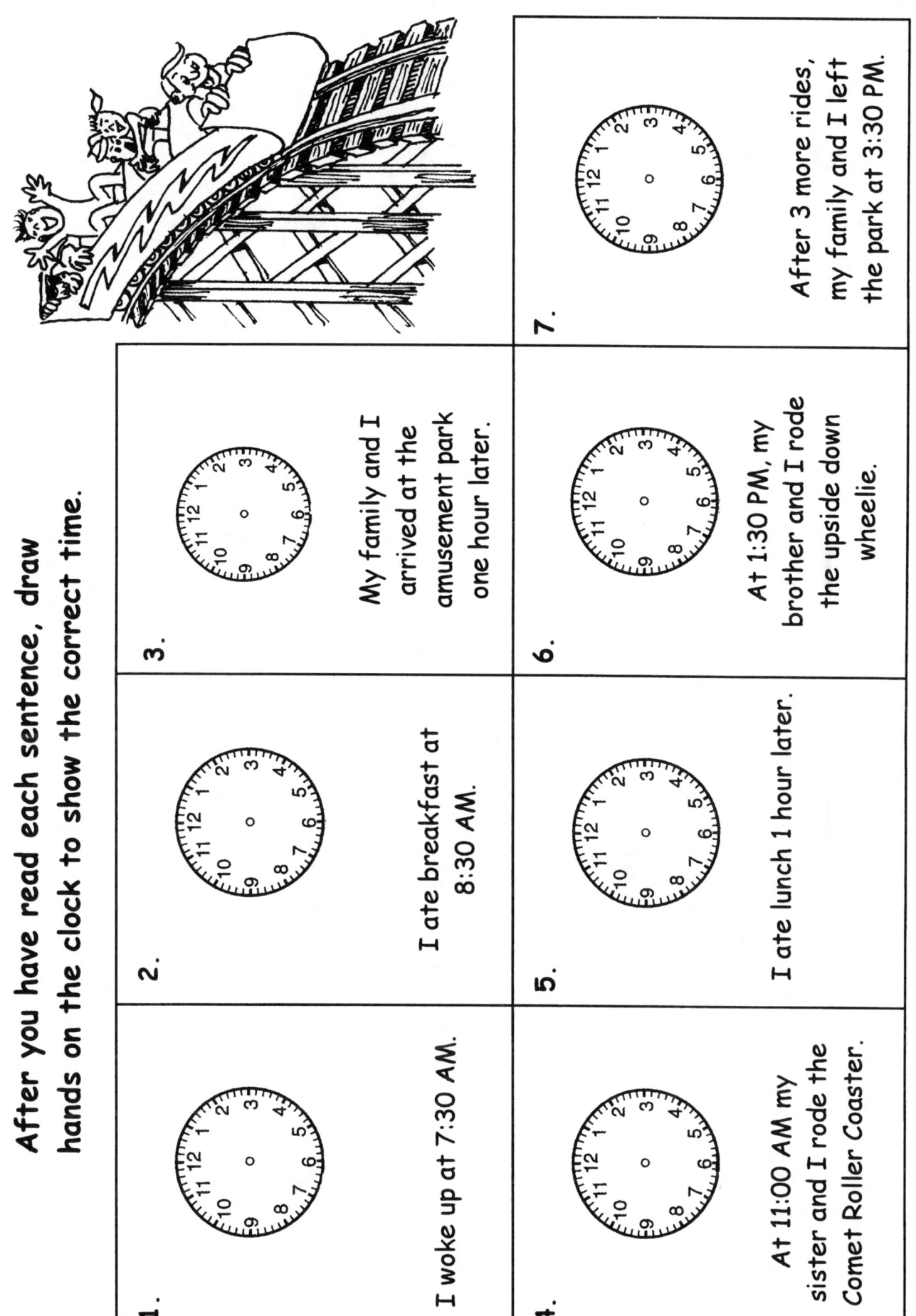

1. I woke up at 7:30 AM.

2. I ate breakfast at 8:30 AM.

3. My family and I arrived at the amusement park one hour later.

4. At 11:00 AM my sister and I rode the Comet Roller Coaster.

5. I ate lunch 1 hour later.

6. At 1:30 PM, my brother and I rode the upside down wheelie.

7. After 3 more rides, my family and I left the park at 3:30 PM.

©2004 by Incentive Publications, Inc., Nashville, TN.

Miss Froggy's Race to Beat the Clock!

Miss Froggy's alarm clock goes off at 7:00 A.M. She needs to be at her flower shop by 9:00 A.M. Will she make it on time? To find out, read what Miss Froggy must do before she goes to work. Look at the time each task takes. Then fill in the time on the chart she will complete each task. Three have been done for you.

Miss Froggy needs to...	It takes her...	Record time she finishes.
1. Brush her teeth	5 minutes	7:05
2. Wash her face	5 minutes	
3. Eat breakfast	20 minutes	
4. Clean up kitchen	10 minutes	
5. Get dressed	10 minutes	7:50
6. Put on makeup	10 minutes	
7. Brush and style hair	10 minutes	
8. Prepare her lunch	15 minutes	8:25
9. Drive to her flower shop	20 minutes	

1. How many total minutes does it take Miss Froggy to make it to her shop after her alarm clock goes off? _____
2. Will she get there on time? _____ Tell why. _____

_____

Planning a schedule

NAME_____

8

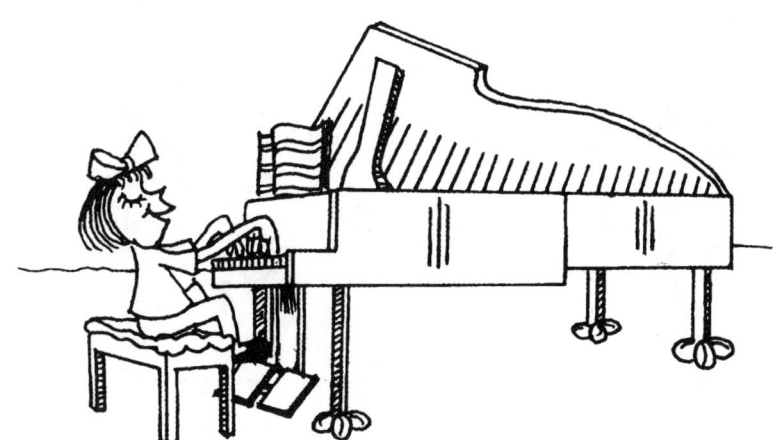

Planning My Own Schedule

Plan your own schedule for one afternoon and evening during the school week.

The following activities must be part of your schedule.

½ hour – homework   ½ hour – dinner   ½ hour – reading

My Afternoon Schedule

3:00	
3:30	
4:00	
4:30	
5:00	
5:30	
6:00	
6:30	
7:00	
7:30	

©2004 by Incentive Publications, Inc., Nashville, TN.

NAME _____

Word problems involving time

It Happened in the Barnyard!

After you have read each sentence, write the time on the line and then draw hands on the clock to show the correct time.

1. Freddy, the rooster, started crowing at 5:30 AM and woke up the entire barnyard. What time was it? _____

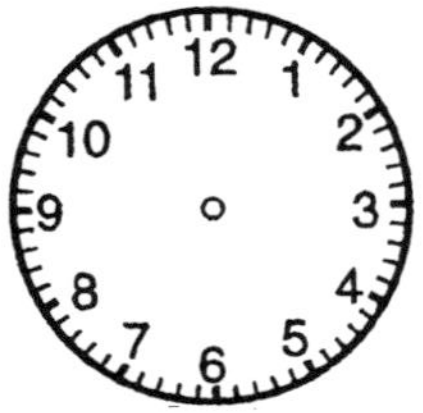

2. Pansy, the pig, was first in line for her food 30 minutes later. What time was it? _____

3. Farmer Fred came 15 minutes later and fed all the animals. What time was it? _____

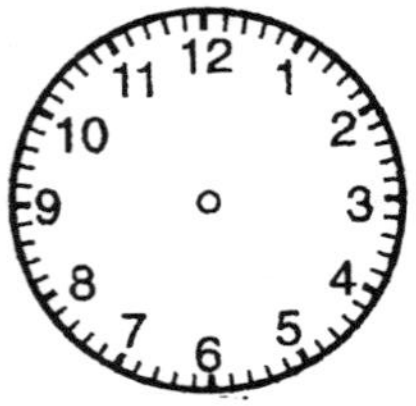

4. Maisy, the cow, was ready to be milked at 15 minutes before 8:30. What time was it? _____

5. Marty, the mule, was hitched up 30 minutes after 10:00 AM. What time was it? _____

6. The geese went for a swim at 12:00 PM and got out of the water 45 minutes later. What time was it? _____

7. Henrietta started sitting on her eggs at 1:00 PM. She finished in 2 hours. What time was it? _____

8. Sadie, the spider, started working on her web at 3:00 PM. She finished 2 hours later. What time was it? _____

©2004 by Incentive Publications, Inc., Nashville, TN.

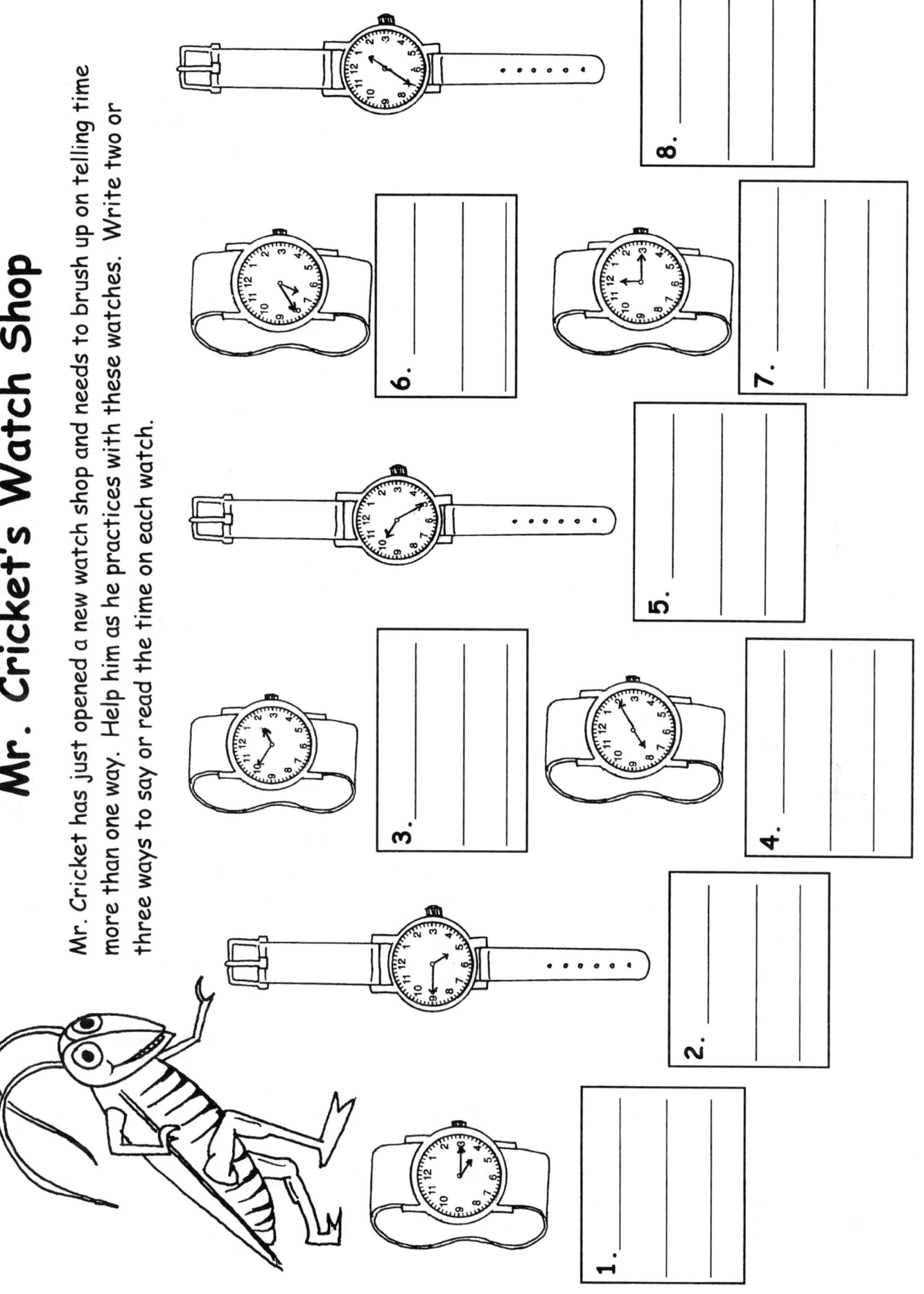

11 Playing MATHO to reinforce math skills

INSTRUCTIONS FOR MATHO

This game is played in the same manner as **BINGO**. The students need a board such as the one on the accompanying page. You, as the teacher, will need one master list of all the possibilities that the students may have on their boards. Several rounds of **MATHO** can be played on one board. For example, a ✔, X, Z, 0, ☺, etc. can be used for different rounds. Some possibilities of games are: (1) regular **MATHO**, which can be won by 4 in a row either horizontally, vertically, diagonally; (2) 4 corners; (3) postage stamp; (4) picture frame; and (5) big "X".

When a student has achieved marking the squares that needed to be called in order to win, he/she yells **MATHO**. The student then brings his paper to you so it can be checked for accuracy. Winners can be awarded prizes such as a new pencil, eraser, computer pass, media center pass, sticker, treasure box toy, or piece of candy from the candy jar, etc.

MATHO can be used to practice many math skills such as whole number facts, money, time, measurement, geometry terms, fractional equivalents, or fraction names. Once you use **MATHO** you will discover many other areas of the curriculum in which it can be used: spelling words, and science or social studies vocabulary review.

On the accompanying page, students will have an opportunity to practice their skills in telling and reading time. One suggestion is to show students a time on a clock or on the overhead and ask them to mark the corresponding time on their **MATHO** boards. Another is to have students choose times from the master list and then the teacher says the time in a different way. For example, the student has 3:15 on his/her board and the teacher says," 15 minutes after 3. Below is a master list that can be used for playing **MATHO**. This list can be put on an overhead transparency, white board, chart, etc. and students choose any 16 of the times to place on their individual boards.

TEACHER'S MASTER LIST

7:30	12:30	4:15	12:20	5:35	2:45
4:45	8:05	2:30	9:25	11:20	3:55
8:00	10:10	3:05	1:15	9:40	5:10
6:35	10:45	2:25	3:50	1:50	6:30

©2004 by Incentive Publications, Inc., Nashville, TN.

Playing MATHO to reinforce math skills   NAME_____   12

MATHO

Choose 16 items from your teacher's list and write them in the boxes. You are now ready to play **MATHO!**

©2004 by Incentive Publications, Inc., Nashville, TN.

Putting the months of the year in order

NAME _____

Putting Them In the Right Order!

November January March June
July February April August
May September October December

Write the 12 months in order. The first one has been done for you.

1. January
2. _____
3. _____
4. _____
5. _____
6. _____
7. _____
8. _____
9. _____
10. _____
11. _____
12. _____

What month do you have a birthday? _____

What month other than your birthday month is your favorite? _____

©2004 by Incentive Publications, Inc., Nashville, TN.

Reading a family time line NAME_____ 14

READING A TIME LINE

Below is an example of a time line. You can see many special days have been shown on the time line such as births, moving, special celebrations, and vacations. On the next page, you will get to make a family time line of your life. Your teacher and your parents can help you.

Maddie's TIME LINE

1996 — I was born in Atlanta, Georgia.

1997 — I moved to St. Louis, Missouri.

1998 — My brother, Michael, was born.

1999 — I started preschool.

2000 — We went to Disney World.

2001 — My grandparents came from Italy to visit.

2002 — I started elementary school.

2003 — I moved to San Francisco, California.

©2004 by Incentive Publications, Inc., Nashville, TN.

15 Constructing a time line   NAME_____

MY TIME LINE

My Picture

Decide on seven events that have been very important in your life. Write the years in the boxes on the left and the events on the lines on the right. You may want to draw a picture of yourself or your family, or you may want to glue a photo of you in the picture box.

Share your time line with your friends. Did they have any events that were the same as the ones on your time line?

©2004 by Incentive Publications, Inc., Nashville, TN.

Reading a time line

NAME_____ 16

Important Inventions in the Last 200 Years

```
         A
<--|--|--|--|--|•-|--|--|--|--|--|--|--|--|--|--|-->
 1830 1840 1850 1860 1870 1880 1890 1900 1910 1920 1930 1940 1950 1960 1970 1980 1990
```

1. Write the letter for each event above its date on the time line. The first one has been done for you.

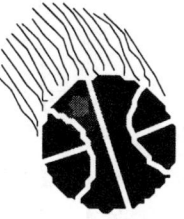

A	Telephone (1876)	H	Barbie Doll (1959)
B	Basketball (1891)	I	Television (1924)
C	Typewriter (1867)	J	Airplane (1903)
D	CD Player (1984)	K	Personal Computer (1973)
E	Blue Jeans (1850)	L	Air Bags (1973)
F	Coca-Cola (1886)	M	Camera (1888)
G	Movie Camera (1894)	N	Radio (1906)

1. What invention on the chart was invented first? _____

2. How many years between the invention of blue jeans and the personal computer? _____

3. About how many years ago were these things invented:

CD Player _____   Air Bags _____

Coca-Cola _____   Airplane _____

Barbie Doll _____   Basketball _____

4. Which invention on the chart was invented last? _____

5. Which invention on the chart is your favorite? _____

6. Which invention do you think has been the most important? _____ Why? _____
_____

©2004 by Incentive Publications, Inc., Nashville, TN.

17 Using money   NAME_____

Miss Froggy's School Shopping List

School Supplies
Pencil.................................20¢
Eraser.................................25¢
Notebook paper....................40¢
Crayons..............................36¢
Glue...................................31¢
Ruler..................................22¢
Scissors.............................33¢
Construction Paper..............41¢

Miss Froggy has run out of school supplies. Look at how much money is shown in each box. Write it in the space. Circle the school supply to match each amount.

1.	(quarter, dime, quarter)	____¢	notebook paper / construction paper
2.	(dime, dime, penny, penny)	____¢	ruler / scissors
3.	(dime, dime, nickel, nickel, penny)	____¢	glue / ruler
4.	(dime, quarter, nickel, nickel)	____¢	eraser / construction paper
5.	(quarter, nickel, nickel, penny)	____¢	glue / crayons
6.	(dime, nickel, nickel, nickel)	____¢	pencil / eraser
7.	(quarter, nickel, penny, nickel, penny)	____¢	crayons / scissors

©2004 by Incentive Publications, Inc., Nashville, TN.

Using money

NAME_____

18

LIZZY NEEDS YOUR HELP!

Lizzy, the ladybug, needs your help. She has lost some coins. Help her find which coin is missing and then draw the correct symbol. Use Q for **quarter**, N for **nickel**, D for **dime**, and P for **penny**.

1. 65¢	2. 34¢	3. 62¢
4. $1.41	5. 91¢	6. 47¢
7. 28¢	8. $1.56	9. $2.30

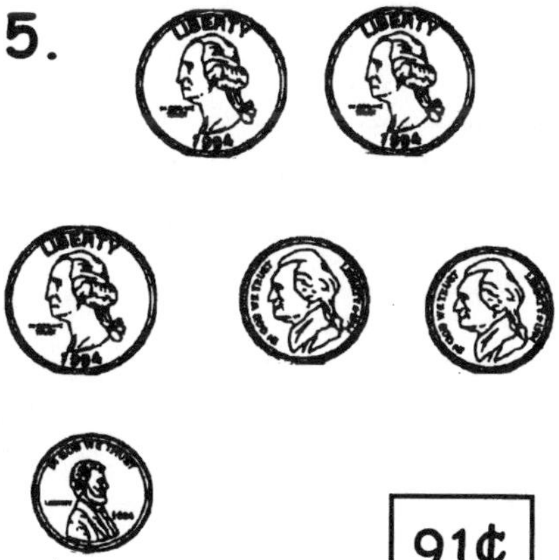

©2004 by Incentive Publications, Inc., Nashville, TN.

19 Sorting coins   NAME_____

Let's Go Shopping!

Choose the fewest coins to buy each of the birthday gifts. Cut out the coins you have chosen and paste the coins in place.

1.	kite 41¢	
2.	jump rope 20¢	
3.	11¢	
4.	checkerboard 50¢	
5.	ball 15¢	
6.	lollipop 5¢	

©2004 by Incentive Publications, Inc., Nashville, TN.

Comparing Costs   NAME_____   20

What's the Best Buy?

Art Supplies					
Watercolors	$4.35	Glue	$1.45	Balloons	$1.49
Popsicle Sticks	$2.25	Scissors	$1.67	Pipe Cleaners	$1.57
Beads	$1.60	Paint Brushes	$2.35	Construction Paper	$1.69

For each problem circle the item that costs more. Subtract to find out how much more. Write your answer in the "answer box". The first one has been done for you.

1. Beads or Construction Paper

1.69
−1.60
―――
 .09

Answer: 9¢

2. Glue or Pipe Cleaners

Answer:

3. Paint Brushes or Balloons

Answer:

4. Watercolors or Scissors

Answer:

5. Glue or Scissors

Answer:

6. Beads or Pipe Cleaners

Answer:

BONUS: Use your calculator and find the total of all the art supplies. Circle the correct answer: $17.42 $20.32 $18.42 $19.22

©2004 by Incentive Publications, Inc., Nashville, TN.

MAKING AN ORGANIZED LIST

Tyler likes all kinds of animals. He has a bunny, two gerbils, a labrador, and two cats. He is at the pet store and sees a toy for his new kitty, Tigger. The toy is 40 cents. He reaches into his pocket and takes out 10 coins. What coins did he use to buy the toy? He may not need to use all of his coins.

Make a list of coins that could be used for Tyler to pay for his toy. One has been done for you.

Quarters	Dimes	Nickels	Pennies	Total number of coins	Value of coins
0	3	1	5	9	40¢

If you could choose a new pet, what pet would you choose? _____

Using money

NAME_____

22

MONEY TALK

Draw coins to solve each riddle. Use Ⓠ for quarter, Ⓓ for dime, Ⓝ for nickel, and Ⓟ for penny.

1. We are 3 coins worth 35¢. What coins are we?		
2. We are 4 coins worth 20¢. What coins are we?	3. We are 5 coins worth 29¢. What coins are we?	4. We are 3 coins worth 30¢. What coins are we?
5. We are 3 coins worth 51¢. What coins are we?	6. We are 4 coins worth 26¢. What coins are we?	7. We are 6 coins worth 52¢. What are we?

©2004 by Incentive Publications, Inc., Nashville, TN.

23 Counting coins   NAME_____

BANANA SPLIT HEAVEN!

How much is shown? Color a scoop when you write the amount.

1.	quarter, dime, dime, quarter	_____ ¢
2.	dime, dime, dime, nickel, penny, penny	_____ ¢
3.	dime, dime, dime, dime, nickel	_____ ¢
4.	dime, dime, dime, nickel, penny	_____ ¢
5.	quarter, quarter, nickel, quarter, nickel	_____ ¢
6.	dime, dime, dime, dime, nickel, nickel, nickel, penny, penny	_____ ¢

©2004 by Incentive Publications, Inc., Nashville, TN.

Using different coins

NAME_____

24

What's in the Treasure Chest?

There are many different coins you can choose to make the total amount. Fill in the blanks. The first one has been done for you.

Total amount	quarters	dimes	nickels	pennies
1. $.18	0	1	1	3
2. $.33	1	0		3
3. $.54		0	0	
4. $.14		1		
5. $.78	3	0	0	
6. $.60	0	4		0
7. $.22		1		2
8. $.38	1		0	3
9. $.25	0		3	0
10. $.65	1	2		0
11. $.11	0	0	1	
12. $.42	1	1		
13. $1.21	4	1		
14. $1.34	4	2		

©2004 by Incentive Publications, Inc., Nashville, TN.

25 Adding using a variety of coins

NAME_____

Shapes, Shapes, and More Shapes

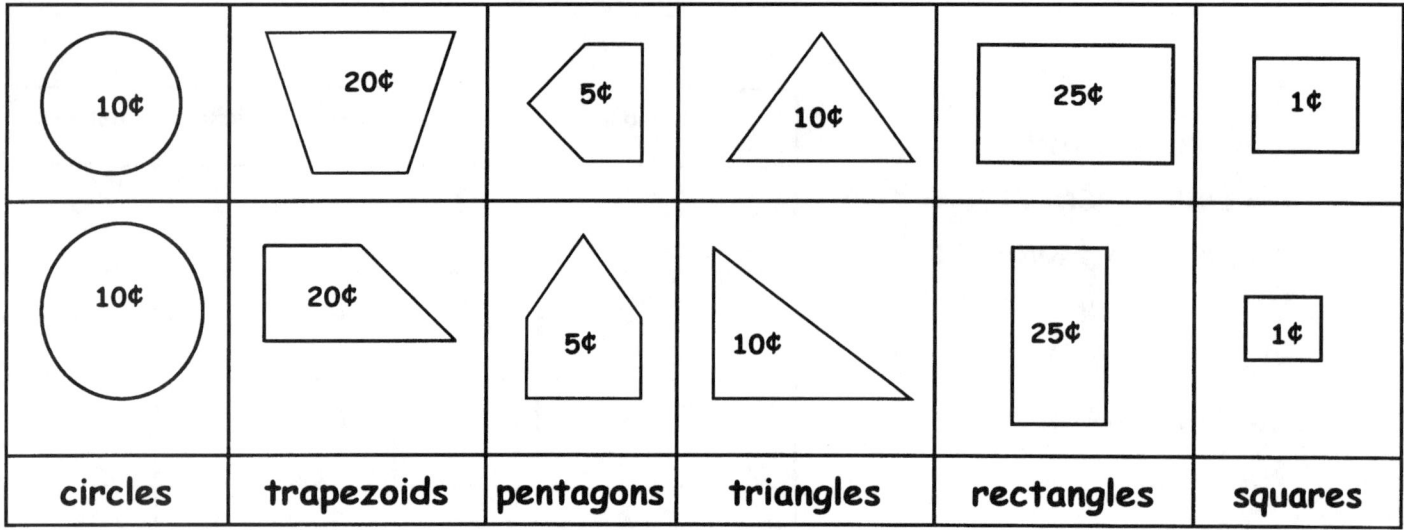

Write the total value of each of the pictures below. The first one has been done for you.

1.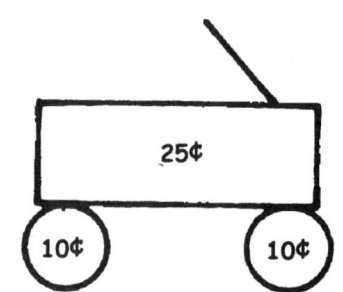

Price: __45¢__

2.

Price: _____

3.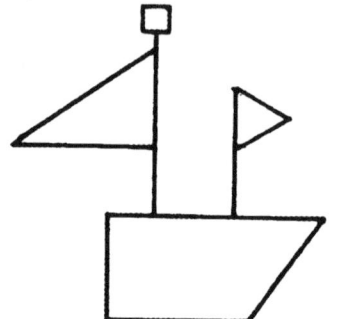

Price: _____

4.

Price: _____

5. Using the shapes above, draw a vehicle that costs between 70¢ and 90¢.

Bonus: Draw another figure on the back of your paper and see if you can make it worth more than $1.00.

©2004 by Incentive Publications, Inc., Nashville, TN.

Adding bills

NAME_____

26

GET ABOARD!

Check out the model train setup below. Find the cost of each part of the set and then find the total cost of the set.

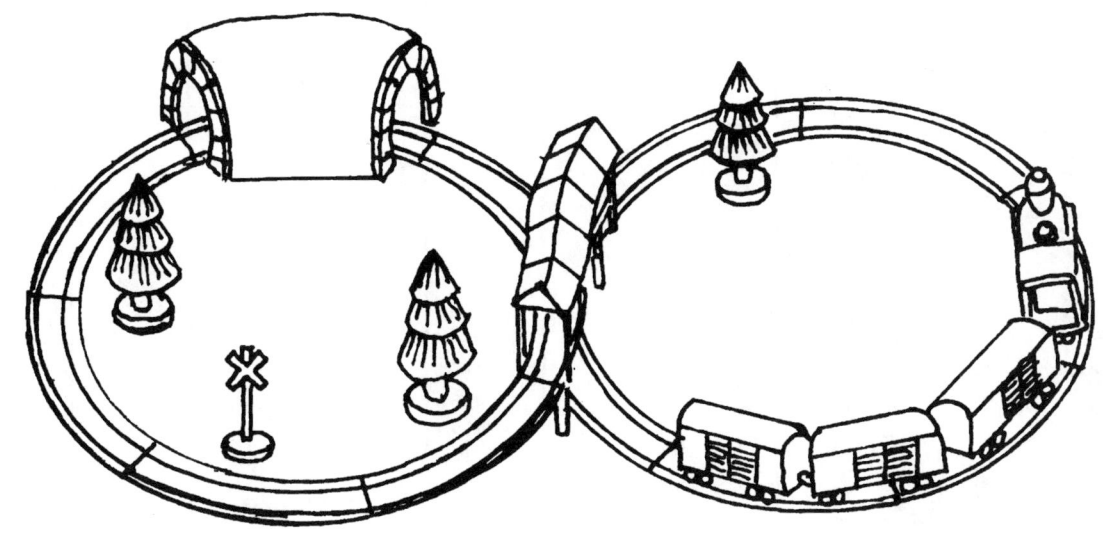

Item	How many?	Cost per Item	Total Cost
1. Train Engine		$5.00	
2. Train Boxcars		$3.00	
3. Pieces of Track	13	$1.00	$13.00
4. Bridge		$5.00	
5. Tunnel		$4.00	
6. Trees		$2.00	
7. Sign		$1.00	

Total cost of Train set _____

Bonus: You got $50.00 for your birthday. Can you buy the set? Yes ____ No ____
If not, how much more money do you need? _____

©2004 by Incentive Publications, Inc., Nashville, TN.

NAME _____

You've Won a Shopping Spree!

Pretend you have won $500.00. Use your calculator and one of your favorite catalogs to shop for your items. Write your choices on the lines below. Try to get as close to $500.00 without spending more than $500.00.

Name of Item	Price	Name of Item	Price
		Total Price	

Using a calculator to shop in a catalog

©2004 by Incentive Publications, Inc., Nashville, TN.

Using the calculator for checking accuracy NAME _____   28

The Detective's Assistant

DIRECTIONS: Sherlock needs your help in checking his work. If Sherlock's answer is correct, draw a ☺ next to it. If his answer is incorrect, draw a ☹ to it.

1. $.90 − .61 $.29 ☺	2. $3.20 + 7.99 $10.61	3. $4.20 6.82 + 9.01 $ 20.03
4. $39.81 − 14.96 $24.85	5. $8.36 .84 + 1.30 $10.50	6. $3.00 − .88 $2.02
7. $8.06 − 2.43 $ 6.63	8. $10.00 − 6.74 $ 3.26	9. $ .16 .39 + 1.38 $2.93

Do you have **5** happy faces and **4** sad faces?
Yes ___ No ___   If not, check your work.
Sherlock wants to thank you for your help.

©2004 by Incentive Publications, Inc., Nashville, TN.

What nursery rhyme do ghosts like best?

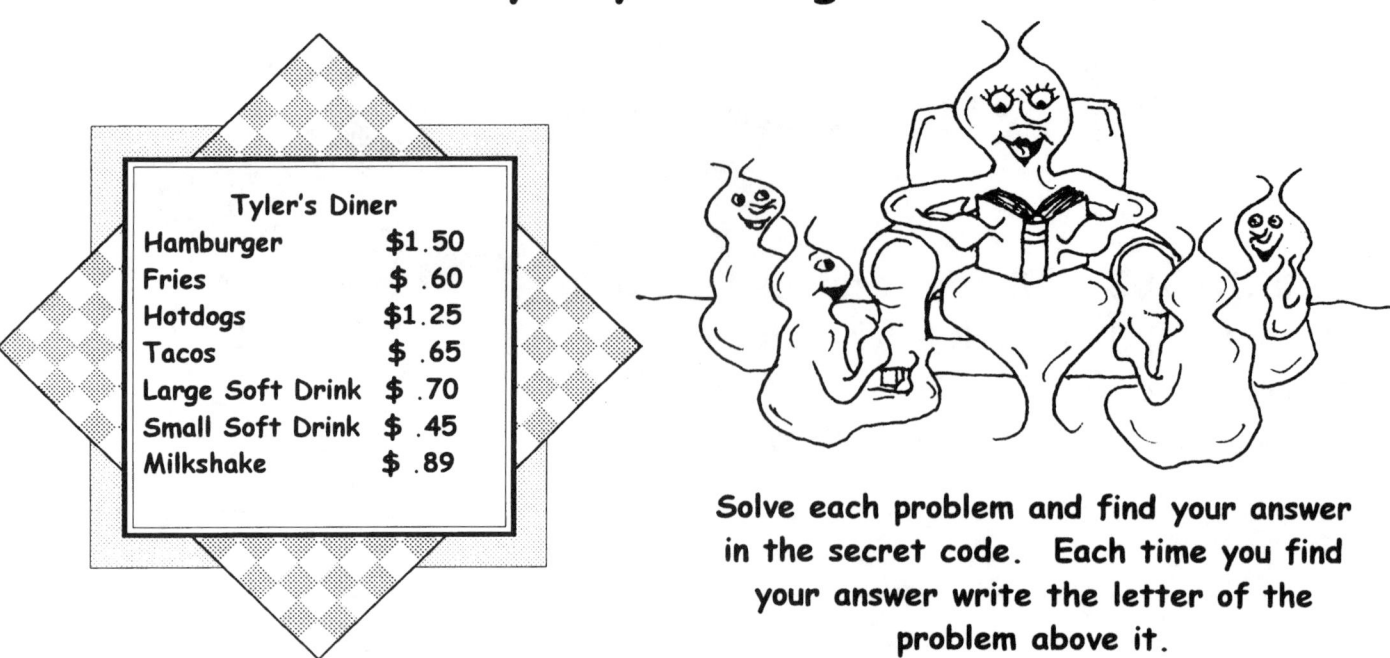

Tyler's Diner
Hamburger $1.50
Fries $ .60
Hotdogs $1.25
Tacos $ .65
Large Soft Drink $ .70
Small Soft Drink $ .45
Milkshake $ .89

Solve each problem and find your answer in the secret code. Each time you find your answer write the letter of the problem above it.

1. Drew orders a hamburger with fries and a small soft drink. How much does he spend? _____ = O

2. Maddie orders 1 hotdog and a large soft drink. How much does she spend? _____ = I

3. How much more is a large soft drink than a small soft drink? _____ = T

4. Clarissa gave the cashier $5.00 for 2 tacos and a large soft drink. How much money does she get back? _____ = E

5. Taylor purchased 3 milkshakes for himself and his friends. How much did he spend? _____ = B

6. Meredith bought 1 taco, fries and a small drink. How much did she spend? _____ = P

7. If you bought one of each item, what would be the total cost? _____ = L

____ ____ ____ ____ ____ ____
$6.04 $1.95 $.25 $.25 $6.04 $3.00

____ ____ ____ ____ ____ ____ ____
$2.67 $2.55 $2.55 $1.70 $3.00 $3.00 $1.70

Adding with money   NAME_____   30

Why couldn't the flower ride her bike to school?

Solve each problem. Find your answer in the secret code. Write the letter of the problem above it.

1. $.23 +.12 =W	2. $.40 +.03 =R	3. $ .36 + 1.23 = E	4. $2.34 + 1.22 =T	5. $.55 + .23 =O
6. $.45 + .29 = N	7. $3.21 + 1.36 = A	8. $.67 + .09 = B	9. $1.86 + .35 = I	10. $ .78 + 1.23 =L
11. $.56 + .32 = S	12. $5.34 + 2.41 =K	13. $1.67 + 2.34 =P	GOOD FOR YOU!	

___ ___ ___ ___ ___ ___ ___ ___ ___
$2.21 $3.56 $.88 $4.01 $1.59 $.3.56 $4.57 $2.01 $.88

___ ___ ___ ___ ___ ___ ___ ___ ___
$.35 $1.59 $.43 $1.59 $.76 $.43 $.78 $7.75 $1.59 $.74

©2004 by Incentive Publications, Inc., Nashville, TN.

31 Subtracting with money

NAME_____

What happened when 2 frogs went after the same bug?

Solve each problem. Find the answer in the secret code. Write the letter of the problem above it.

1. $.47 − .23 =Y	2. $.68 − .34 =A	3. $1.23 − .34 =U	4. $.31 − .10 =M	5. $4.31 − 2.20 =D
6. $.99 − .21 = B	7. $.56 − .45 =N	8. $1.75 − .34 =H	9. $.83 − .27 =E	10. $1.37 − .39 =O
11. $.68 − .35 =G	12. $1.23 − 1.05 =I	13. $.92 − .26 =C	14. $5.23 − 2.56 =T	

Way to go!!!

___ ___ ___ ___   ___ ___ ___ ___ ___ ___
$2.67 $1.41 $.56 $.24 $.78 $.56 $.66 $.34 $.21 $.56

                              −
___ ___ ___ ___ ___ ___ ___ ___ ___ ___
$2.67 $.98 $.11 $.33 $.89 $.56 $2.67 $.18 $.56 $2.11

©2004 by Incentive Publications, Inc., Nashville, TN.

Adding and subtracting with money

NAME_____

What did the out-of-work skunk say?

Solve each problem.
Find your answer in the secret code.
Write the letter of the problem above it.

1. $.89 + .34 = V	2. $1.23 − .45 = E	3. $.98 + .23 = A	4. $1.45 + 2.45 = T
5. $1.31 − .45 = D	6. $2.45 − 1.34 = I	7. $.98 − .33 = C	8. $4.35 + 2.36 = H
9. $1.52 − 1.04 = N	10. $1.41 + 4.54 = K	11. $4.56 − 2.39 = S	12. $6.78 + 1.54 = O

,

___ ___ ___ ___ ___   ___ ___ ___ ___
$1.11 $.86 $8.32 $.48 $3.90 $6.71 $1.21 $1.23 $.78

___ ___ ___ ___ ___   ___ ___ ___
$1.21 $2.17 $.65 $.78 $.48 $3.90 $1.21 $.48 $.86

___ ___ ___ ___ ___ ___ ___ ___
$1.11 $3.90 $2.17 $3.90 $1.11 $.48 $5.95 $2.17

©2004 by Incentive Publications, Inc., Nashville, TN.

33 Thinking of many ways to make a dollar

NAME_____

What's in a Dollar?

Think of many ways to make a dollar. One has been done for you. Try to think of at least 6-10 other ways.

1.	10¢ + 25¢ + 5¢ + 40¢ + 20¢ = $1.00
2.	
3.	
4.	
5.	
6.	
7.	
8.	
9.	
10.	

©2004 by Incentive Publications, Inc., Nashville, TN.

Writing word problems with money

NAME _____

MONEY PROBLEMS!

Write three story problems using money.
Ask a friend to solve your problems.

1.	2.	3.

©2004 by Incentive Publications, Inc., Nashville, TN.

Standard and Metric Measurement Skill Builders

©2004 by Incentive Publications, Inc., Nashville, TN.

Using measurement charts NAME_____ 36

Measurement Helper

Length

Kilometer: 1000 meters
Meter: 100 cm or 1000 millimeters
Millimeter: $\frac{1}{1000}$ of a meter
Foot: 12 inches
Yard: 3 feet or 36 inches
Mile: 5,280 feet or 1,760 yards
10 centimeters is about 4 inches

Time

Millennium = 1000 years
Century = 100 years
Decade = 10 years
Year: 365 or 366 days or about 52 weeks or 12 months
Month: 28, 29, 30, or 31 days
Week: 7 days
Day: 24 hours
Hour: 60 minutes
Minute: 60 seconds

Weight

Kilogram: 1000 grams
Gram = 1000 milligrams
Milligram = $\frac{1}{1000}$ of a gram
Pound = 16 ounces
Ton = 2000 pounds
1 ounce is about 30 grams

Mathematical Symbols

< is less than
> is greater than
= is equal to/is the same as
+ to add
− to subtract
× to multiply
÷ to divide

Temperature

0° Celsius (C) = Water freezes
100° Celsius (C) = Water boils
27° Celsius (C) = A comfortable day
32° Fahrenheit (F) = Water freezes
212° Fahrenheit (F) = Water boils
98.6° Fahrenheit (F) = Normal body temperature
75° Fahrenheit (F) = Normal spring day

Money

= 1¢ or $0.01   P
= 5¢ or $0.05   N
= 10¢ or $0.10 D
= 25¢ or $0.25 Q
= 100¢ or $1.00 $1

Abbreviations

kilometers = km
meters = m
centimeters = cm
millimeters = mm
miles = mi
feet = ft
inches = in
liters = l
milliliters = ml
yards = yd
ounces = oz
pounds = lb
tons = t
kilograms = kg
grams = g
milligrams = mg

©2004 by Incentive Publications, Inc., Nashville, TN.

Solving measurement problems

NAME _____

Where do rabbits go when they get married?

Use your measurement tables and solve the problems. Find your answer in the secret code. Write the answer of the problem above it.

1. 4000 pounds = _____ tons (H)
2. $1.00 = _____ pennies (O)
3. 2 days = _____ hours (T)
4. The abbreviation for pounds is _____ (U)
5. 240 seconds = _____ minutes (I)
6. 5000 grams = _____ kilograms (N)
7. 300 centimeters = _____ meters (Y)
8. 20 cm is about how many inches _____ (R)
9. 52 weeks = _____ year/years (M)
10. 4 weeks = _____ days (E)
11. The abbreviation for ounces is _____ (B)

___ ___ ___ ___ ___ ___ ___ ___ ___
100   5   48   2   28   4   8

___ ___ ___ ___ ___ ___
5   5   3   1   100   5

___ ___
oz   lb

©2004 by Incentive Publications, Inc., Nashville, TN.

Learning liquid measurement

NAME_____

38

MEASUREMENT CLOWN

Materials: construction paper, scissors, ruler, markers, glue

Directions: Using the drawing of Measurement Clown, ask students to use different colored construction paper for each measurement, and then cut and glue the correlating pieces of construction paper onto the drawing, or make a larger scale version of the Measurement Clown.

> An easy way to remember liquid measurement from the smallest to the largest is to look at the size of the words! The smallest measurement, the cup, has 3 letters, which makes it smaller than a pint (4 letters).
>
> c-u-p = 3 letters   q-u-a-r-t = 5 letters
> p-i-n-t = 4 letters   g-a-l-l-o-n = 6 letters

Below are given the corresponding standard measurement pieces that are found on the Measurement Clown.

☞ cup
☞ pint
☞ quart
☞ gallon

Use your Measurement Clown to solve these problems:

1.   4 pints = _____ quarts
2.   4 quarts = _____ gallon/gallons
3.   2 gallons = _____ quarts
4.   4 cups = _____ pints

5.   2 cups = _____ pint/pints
6.   8 pints = _____ quarts
7.   1 gallon = _____ pints
8.   4 pints = _____ cups

©2004 by Incentive Publications, Inc., Nashville, TN.

NAME _____

"Mary, Mary, Mary, How Does Your Garden Grow?"

First, plant a seed. Second, measure your plant with an inch ruler at the end of each week. Use the chart below to write down how much your plant has grown.

Height of plant	Week 1	Week 2	Week 3	Week 4	Week 5

"Mary, Mary quite contrary, how does your garden grow?"
"With silver bells and cockle shells and pretty maids all in a row."

Bonus: Were there any surprises about the plant you grew? Share those with a classmate.

You may want to color the garden and the young girl who is watering the plants.

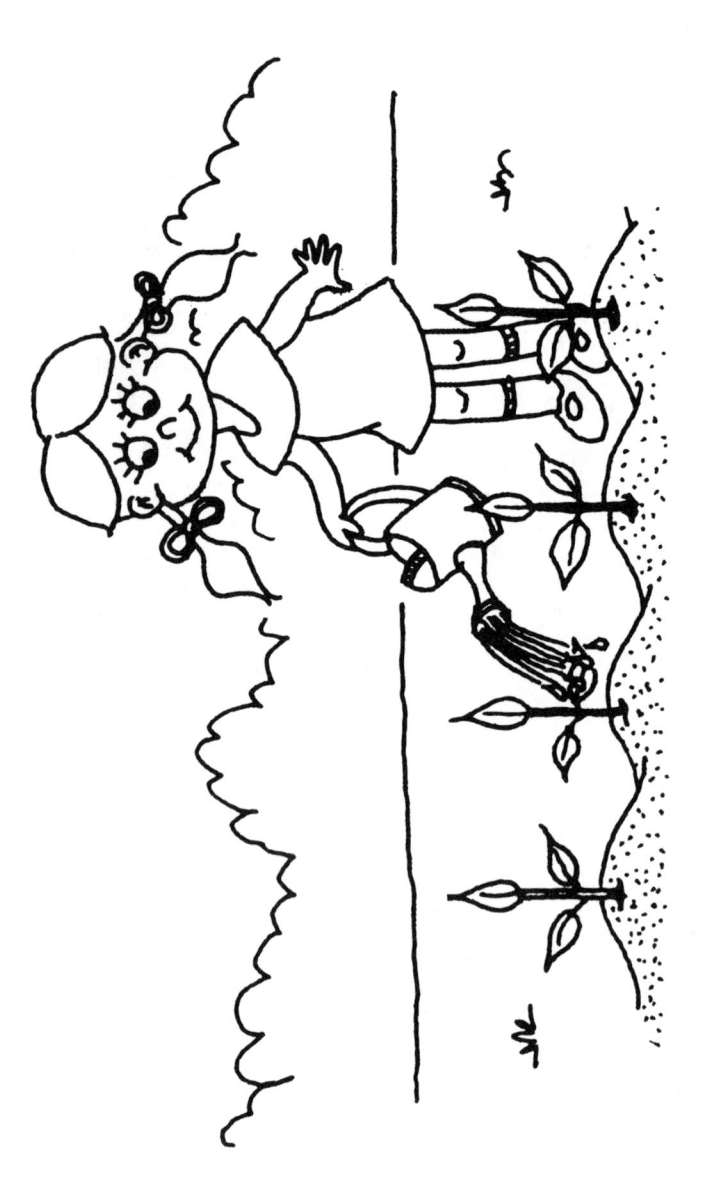

Connecting standard measurement to the real world

©2004 by Incentive Publications, Inc., Nashville, TN.

NAME _____

Where do fish keep their money?

Capacity Measurement
1 pint = 2 cups
1 quart = 2 pints
1 gallon = 4 quarts
1 pound = 16 ounces
1 ton = 2,000 lbs

Solve each problem. Find your answer in the secret code. Each time you find your answer, write the letter of the problem above it.

1. 4 quarts = _____ gallon/s (A)

2. 32 ounces = _____ pounds (N)

3. 6000 pounds = _____ tons (V)

4. 8 cups = _____ pints (I)

5. 2 gallons = _____ quarts (K)

6. 3 quarts = _____ pints (B)

7. 2 tons = _____ pounds (E)

8. 2 pounds = _____ ounces (R)

__ __ __ __ __ __ __ __ __ __
4   2   1   32 4   3   4,000 32 6 1 2 8

Solving problems using standard measurement

©2004 by Incentive Publications, Inc., Nashville, TN.

41 Solving problems using standard measurement

NAME_____

Why does a banana wear suntan lotion?

Length Measurement

1 foot = 12 inches

1 yard = 36 inches (or 3 feet)

1 mile = 5,280 feet

Solve each problem. Find your answer in the secret code. Each time you find the answer, write the letter of the problem above it.

1. 1 foot = _____ inches (D)
2. 1 yard = _____ feet (O)
3. ½ foot = _____ inches (E)
4. 5,280 feet = _____ mile/s (L)
5. 1½ feet = _____ inches (T)
6. 15 feet = _____ yards (S)
7. 3 feet = _____ inches (N)
8. 10,560 feet = _____ miles (I)
9. 3 yards = _____ feet (P)

__ __   __ __
5 3   2 18

__ __ __ __ __ , __ __ __ __ __
12 3 6 5 36   18 9 6 6 1

©2004 by Incentive Publications, Inc., Nashville, TN.

Solving problems using standard measurement
NAME_____ 42

Why is the basketball player such a messy eater?

Time Measurement
1 minute = 60 seconds
1 hour = 60 minutes
1 day = 24 hours
1 week = 7 days
1 year = 12 months
1 year = 52 weeks
1 year = 365 days
1 century = 100 years

Solve each problem. Find your answer in the secret code. Each time you find your answer, write the letter of the problem above it.

1. 14 days = _____ weeks (O)
2. 2 minutes = _____ seconds (T)
3. 1 year = _____ weeks (E)
4. 3 hours = _____ minutes (V)
5. 2 centuries = _____ years (H)
6. 365 days = _____ year/s (I)
7. ½ century = _____ years (B)
8. 3 weeks = _____ days (S)
9. 240 minutes = _____ hours (L)
10. 500 years = _____ centuries (P)
11. 70 days = _____ weeks (A)
12. 3 years = _____ months (R)
13. 2 years = _____ days (D)
14. ½ hour = _____ minutes (C)

___ ___ ___   ___ ___ ___   ___ ___ ___ ___ ___   ___ ___ ___
21 200 52   730 36 1   50 50 4 52 21   10 4 4

___ ___ ___ ___   ___ ___ ___ ___   ___ ___ ___ ___ ___
2 180 52 36   120 200 52   5 4 10 30 52

©2004 by Incentive Publications, Inc., Nashville, TN.

43 **Using centimeters to measure distances**

NAME_____

It's All About Centimeters!

Measure to the nearest centimeter how far the dog is away from the bone. Write your answer above each bone.

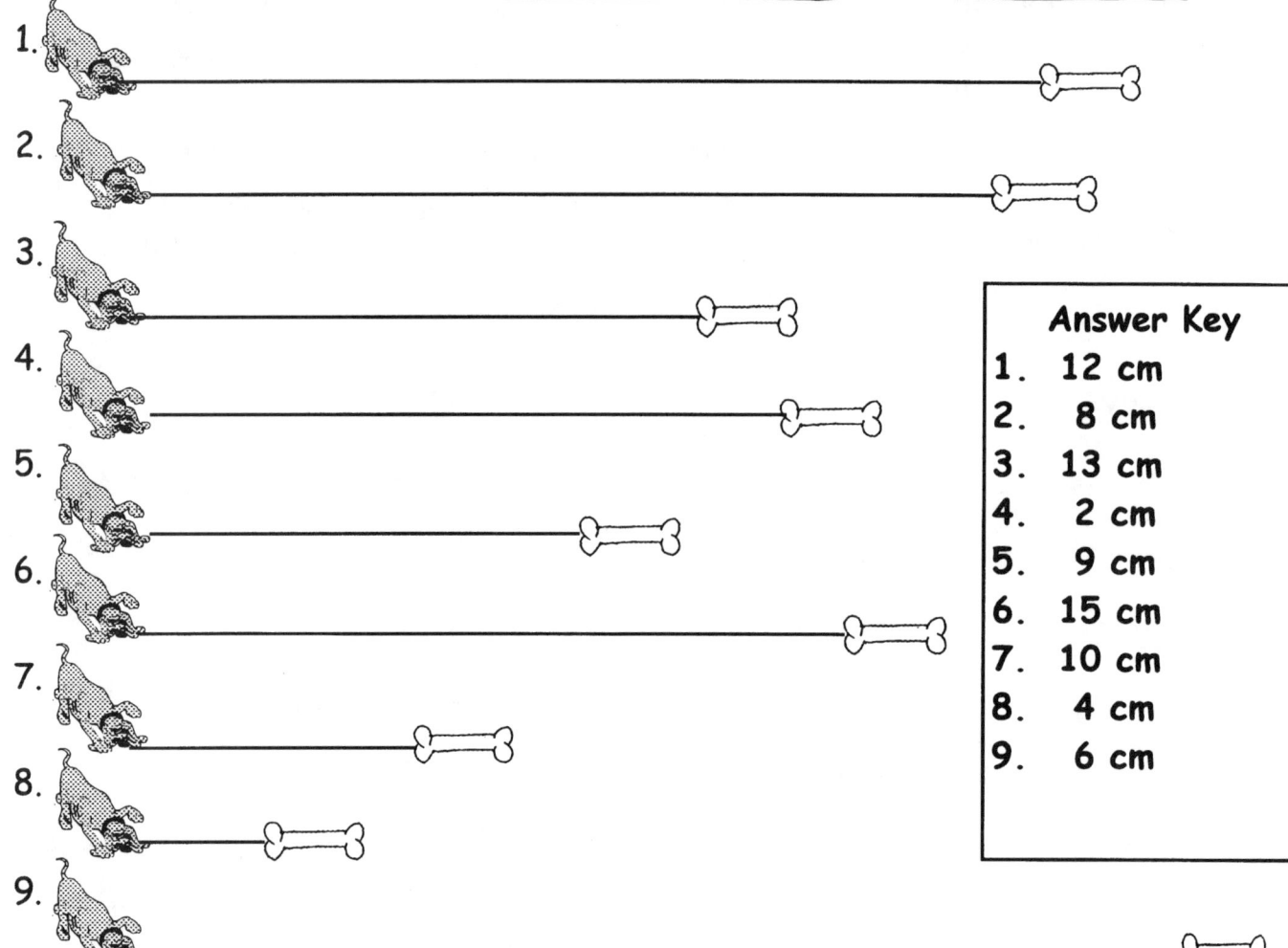

1.
2.
3.
4.
5.
6.
7.
8.
9.

Answer Key
1. 12 cm
2. 8 cm
3. 13 cm
4. 2 cm
5. 9 cm
6. 15 cm
7. 10 cm
8. 4 cm
9. 6 cm

BONUS: Draw three lines on the back of your paper and ask a friend to measure them in centimeters.

©2004 by Incentive Publications, Inc., Nashville, TN.

Measuring real life objects

NAME_____

44

Metric Art Gallery

In the castle outline below draw 5 to 10 objects which measure 30 centimeters or less.

©2004 by Incentive Publications, Inc., Nashville, TN.

45 Using metric measurement   NAME_____

Let's Dig Into Metric Measurement!

Color Code
Meters: green
Grams: orange
Liters: purple

Read each statement and decide what is the correct unit of metric measurement. Use the color code to color each bone.

1. The weight of an orange

2. The height of a ladder

3. The amount of water in an aquarium

4. The height of a door

5. The length of a tennis racquet

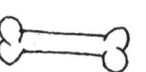

6. The weight of a golf ball

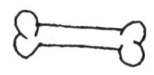

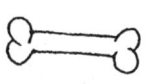

7. The distance across a pool table

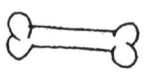

8. The amount of milk in a large carton

9. The weight of a dozen eggs

10. The amount of of water your bathtub will hold

11. The height of a basketball goal

12. The weight of a yoyo

Use the chart to help you.

Use **grams** to measure **weight**.
Use **liters** to measure the **amount of liquid** a container can hold.
Use **meters** to measure **length** or **distance**.

©2004 by Incentive Publications, Inc., Nashville, TN.

Measuring items in the classroom

NAME_____

METRIC SCAVENGER HUNT

Find objects in your classroom that measure the distances listed below.

Find objects that measure about
10 centimeters:

1. _____
2. _____

Find objects that measure about
1 meter:

1. _____
2. _____

Find objects that measure about
50 centimeters:

1. _____
2. _____

Find objects that measure about
60 millimeters:

1. _____
2. _____

Find objects that measure about
30 centimeters:

1. _____
2. _____

Find objects that measure about
15 millimeters

1. _____
2. _____

Find objects that measure about
25 millimeters:

1. _____
2. _____

Find objects that measure about
2 meters:

1. _____
2. _____

Find any four objects and measure those in metrics. Write your objects and how much they measure in the spaces below.

1. _____   3. _____
2. _____   4. _____

©2004 by Incentive Publications, Inc., Nashville, TN.

MAKING THE BEST CHOICES!

Choose the best measurement from the list in each of the three boxes. Write it in the spaces.

Measurement of Length:
meter, millimeter, centimeter, kilometer

1. The length of a baseball bat _____
2. The width of your toenail _____
3. The distance from Atlanta to Los Angeles _____
4. The length of a jellybean _____
5. The length of your bicycle _____
6. The length of a mosquito _____
7. The length of a soccer field _____

Liquid Measurement:
liter, milliliter

1. A small carton of orange juice _____
2. A swimming pool _____
3. A hot tub _____
4. A medicine dropper _____
5. A large fish tank _____
6. A raindrop _____
7. A bathtub _____

Measurement of Weight
gram, milligram, kilogram

1. An apple _____
2. An elephant _____
3. A feather _____
4. A soccer ball _____
5. A small piece of thread _____
6. A bulldozer _____
7. A snowflake _____

©2004 by Incentive Publications, Inc., Nashville, TN.

Using metric measurement

NAME_____ 48

STAR SEARCH in CENTIMETERS

Solve the problem in each of the stars. Write your answers in the stars.

1. Find the length of your math book. ____ cm

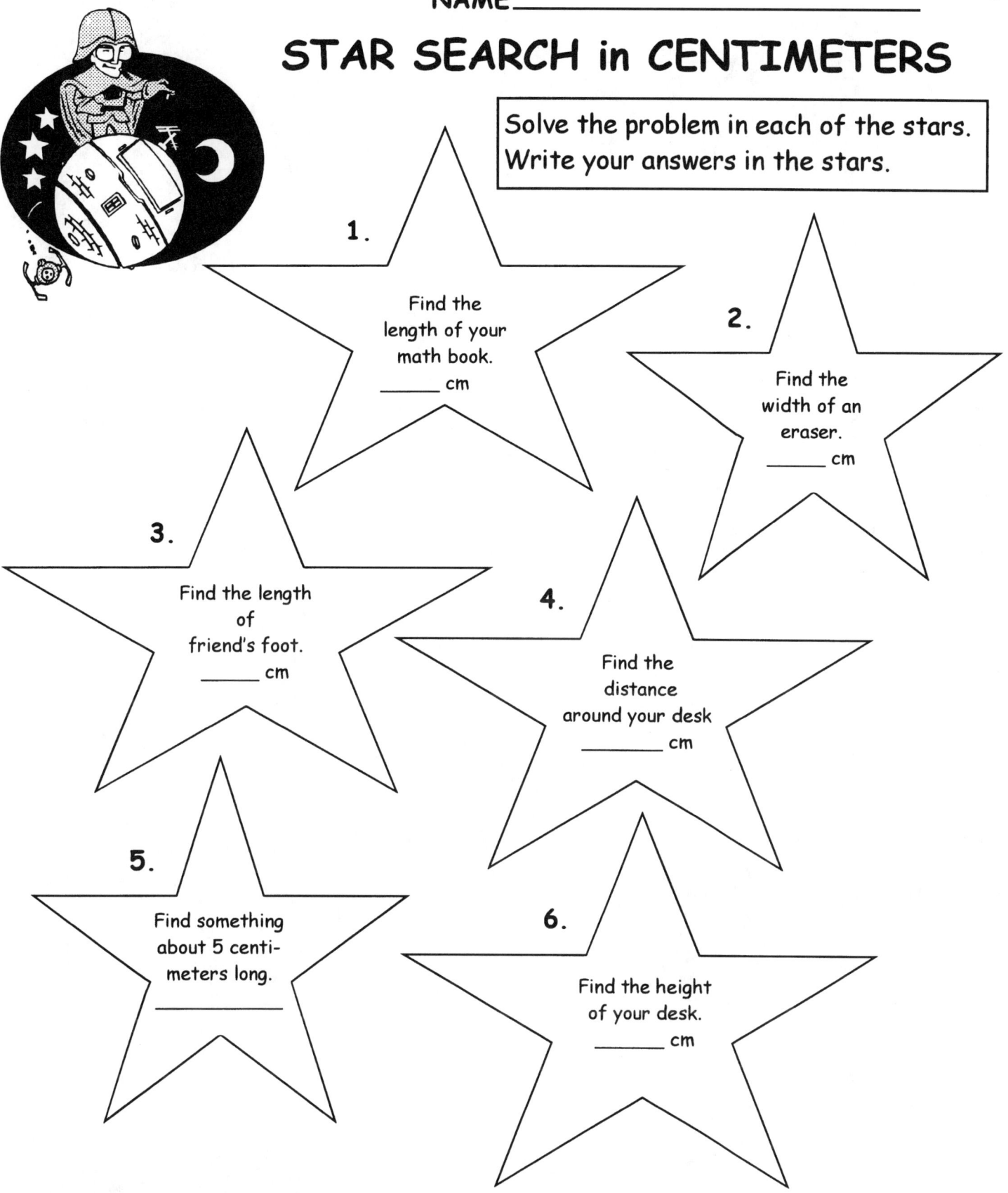

2. Find the width of an eraser. ____ cm

3. Find the length of friend's foot. ____ cm

4. Find the distance around your desk ____ cm

5. Find something about 5 centimeters long. ____

6. Find the height of your desk. ____ cm

Bonus: Find 3 objects that measure about 40 centimeters long:
_____, _____, _____

©2004 by Incentive Publications, Inc., Nashville, TN.

METRIC OCEAN HABITAT

49

Measuring line segments with metric notation

NAME _____

METRIC OCEAN HABITAT

Using the drawing of the metric ocean habitat, measure the following line segments to the nearest centimeter. The first one has been done for you. Find your answer in the answer code and circle it. You may have the same answer for two problems.

Answer Code
A. 17 cm
B. 4 cm
C. 22 cm
D. 7 cm
E. 13 cm
F. 11 cm
G. 18 cm
H. 4 cm
I. 6 cm
J. 16 cm
K. 15 cm
L. 6 cm

1. $\overline{AB}$ = _____ cm
2. $\overline{CD}$ = _____ cm
3. $\overline{EF}$ = _____ cm
4. $\overline{GH}$ = _____ cm
5. $\overline{JK}$ = _____ cm
6. $\overline{GL}$ = _____ cm

7. $\overline{MC}$ = _____ cm
8. $\overline{CL}$ = _____ cm
9. $\overline{EL}$ = _____ cm
10. $\overline{AL}$ = _____ cm
11. $\overline{AI}$ = _____ cm
12. $\overline{LJ}$ = _____ cm

You are the designer! Using the *Metric Ocean Habitat* as a guide, design your own Ocean Habitat. Label 8-10 points with alphabet letters as is done on the *Metric Ocean Habitat*. Ask a friend to measure your line segments.

50

©2004 by Incentive Publications, Inc., Nashville, TN.

MY METRIC HABITAT

Reading Fahrenheit and Celsius Thermometers

NAME _____

What's the Temperature?

Fahrenheit and Celsius Temperature Scales

Fahrenheit (°F) temperatures: -40 -30 -20 -10 0 10 20 30 40 **32** 60 80 100 **98.6** 120 140 160 180 200 **212**

Celsius (°C) temperatures: -40 -30 -20 -10 **0** 10 20 30 **37** 50 60 70 80 90 **100**

Water freezes | Human body temperature | Water boils

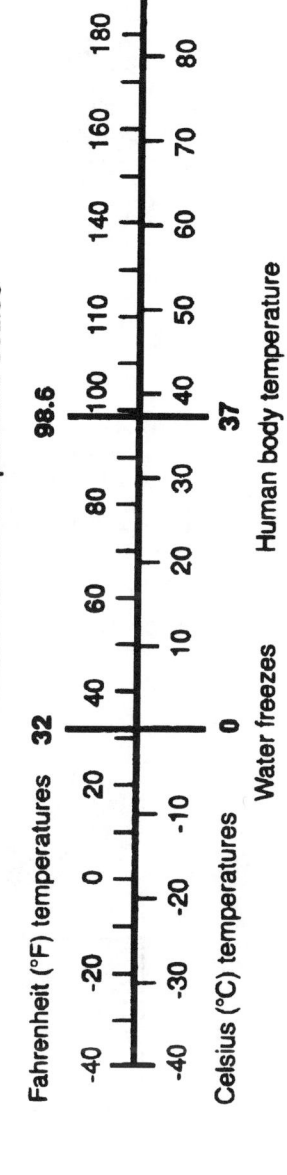

Use the *Celsius* thermometer to answer these questions. What degree is the following?

1. Room temperature _____
2. Water freezes _____
3. Body Temperature _____

Use the *Fahrenheit* thermometer to answer these questions. What degree is the following?

4. Room temperature _____
5. Water freezes _____
6. Body Temperature _____

Which thermometer is most likely being used with these temperatures? Use C for Celsius and F for Fahrenheit.

7. Water starts to freeze at 0° _____
8. Watching TV at 70° _____
9. Camping in the woods at 22° _____
10. Flying a kite at 80° _____
11. Sitting in a hot tub at 90° _____
12. Skateboarding at 27° _____

52

©2004 by Incentive Publications, Inc., Nashville, TN.

Prehistoric Creatures

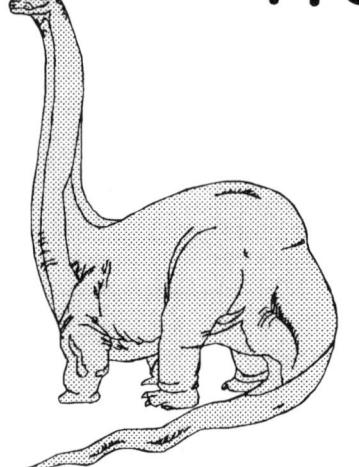

Brontosaurus
Thunder lizard

Triceratops
Weighed as much as nine tons (18,000 lbs)

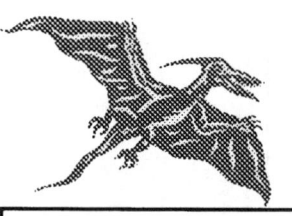

Pterosaurus
Large brain
No feathers

Stegosaurus
Horn-covered, bony plate along its tail and back.

Velociraptor
Ferocious, two-legged predator.
Long-clawed fingers

Tyrannosaurus
Tyrant lizard

Parasaurolophus
Spectacular duckbill crest which made honking noises

Brachiosaurus
One of the largest dinosaurs
Weighed up to 165 tons

Deinonychus
Sickle-like claw on each foot which was used to tear open the stomachs of other dinosaurs

Iguanodon
Spike-like thumb used as a weapon
No teeth

©2004 by Incentive Publications, Inc., Nashville, TN.

Solving word problems

NAME_____

54

WHAT AM I?

Clue: I was one of the last dinosaurs to disappear from the earth. Solve the riddles below using your dinosaur chart. Find your answer in the secret code. Write the letter of the problem above it and solve the question, "What Am I?"

1. I am a 33 foot long dinosaur. I had a spectacular duckbill crest with nasal passages which were used for producing loud honking noises.
   What am I? _____ = I

2. I am the largest horned dinosaur. I resemble a rhinoceros with heavy head shields and horns mounted on my face. I weigh in at nine tons, or 18,000 lbs.
   What am I? _____ = R

3. I am another armored dinosaur. I can reach a length of 20 feet. My skull and brain were very small for such a large animal. I have hard horn plates on my back. I only eat plants and I move rather slowly. What am I? _____ = A

4. I am known as a ferocious two-legged predator and I measured 6 ft. long with a low head, very sharp teeth, and a rather large brain. On each of my feet, my second toe is a large, retractable claw. What am I? _____ = O

5. I am another fierce dinosaur with a sickle-like claw on each foot. This foot is used to tear open the stomachs of other dinosaurs.
   What am I? _____ = E

6. I am a flying dinosaur that looks more like a bat than a bird. I have a large brain compared to those of other birds. My wingspan can reach upwards of 50 ft.
   What am I? _____ = S

7. I am the king of all dinosaurs. I have a huge skull with many sharp razor-like teeth. I use my teeth which are six inches long and 1 inch wide to attack other dinosaurs. I have to eat, you know! What am I? _____ = C

8. I am the first dinosaur discovered in the 19th century. I have bird-like feet and eat only plants. I have no teeth at the front of my jaw, just a bony beak.
   What am I? _____ = T

9. I am one of the largest dinosaurs. I can weigh up to 30 tons which is an unbelievable 60,000 lbs. For that reason, I am known as the "thunder" lizard because I sound like thunder when you hear me coming. What am I? _____ = P

___	___	___	___	___	___	___	___	___	___	___
Iguana-don	Tricera-tops	Parasaur-olophus	Tyranno-saurus	Deinony-chus	Tricera-tops	Stego-saurus	Iguana-don	Veloci-raptor	Bronto-saurus	Ptero-saurus

©2004 by Incentive Publications, Inc., Nashville, TN.

NAME _____

Reading temperatures on a map

The Highs and Lows of It!

U.S. Weather Map: Spring High/Low Temperatures (°F)

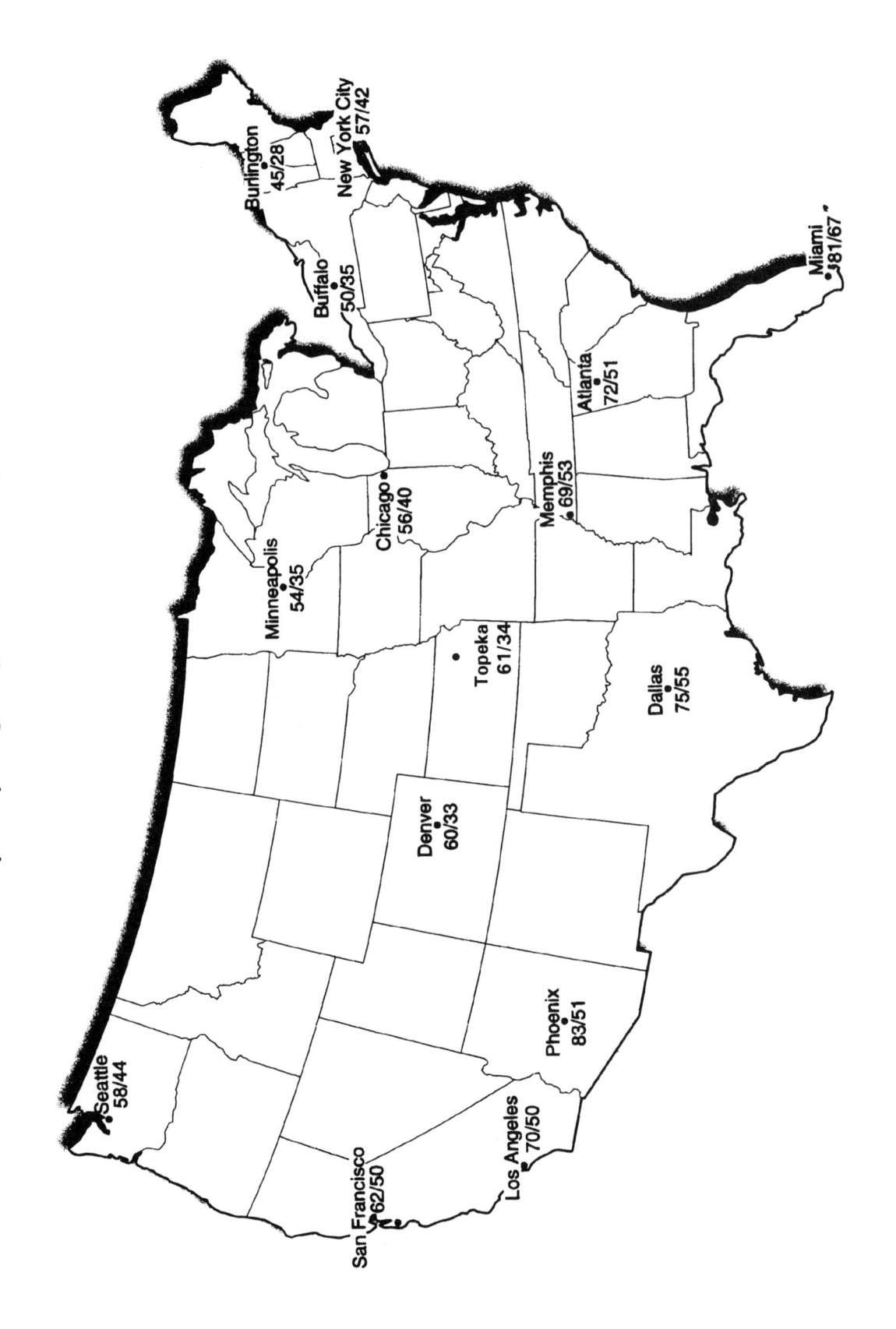

- Burlington 45/28
- New York City 57/42
- Buffalo 50/35
- Miami 81/67
- Atlanta 72/51
- Minneapolis 54/35
- Chicago 56/40
- Memphis 69/53
- Topeka 61/34
- Dallas 75/55
- Denver 60/33
- Phoenix 83/51
- Seattle 58/44
- San Francisco 62/50
- Los Angeles 70/50

©2004 by Incentive Publications, Inc., Nashville, TN.

NAME _____

Using the U.S. Weather Map, answer the questions below.

Complete the chart below. The first one has been done for you.
There are two blank spaces for you to choose two different cities.

1.

City	Warmest Temperature	Coldest Temperature	Difference
Atlanta, Georgia	72°	51°	21°
Minneapolis, Minnesota			
San Francisco, California			
New York City, New York			
Dallas, Texas			

2. Looking at your map answer these questions:
   A. What city has the warmest temperature? _____
   B. What city has the coldest temperature? _____
   C. What is the difference in the two temperatures? _____

3. Name any city/cities you have visited that are on the weather map: _____
4. If you could choose one city to visit, what would be your choice? _____
   Tell why. _____

©2004 by Incentive Publications, Inc., Nashville, TN.

57 Reading a weather map

It's All About the Weather!

The map below shows the number of days of snow or rain during the winter months of January, February, and March. Using the map, answer the questions.

- Barrow — 10 days
- Alaska (U.S.A.)
- Fairbanks — 20 days
- Arctic Bay — 15 days
- Canada
- Vancouver — 55 days
- Winnipeg — 30 days
- Toronto — 40 days
- Halifax — 45 days
- San Francisco — 30 days
- United States
- Chicago — 30 days
- New York City — 35 days
- Death Valley — 5 days
- Los Angeles — 20 days
- Atlanta — 35 days
- Houston — 25 days
- Mexico
- Puerto Rico
- Mexico City — 15 days
- San Juan — 50 days
- Nicaragua
- Managua

©2004 by Incentive Publications, Inc., Nashville, TN.

It's All About the Weather – Page 2

Look at the map and write the number of days it snows or rains in the winter months in each place on the chart below.

	Days of Snow or Rain		Days of Snow or Rain
Toronto, Canada	_____	Fairbanks, Alaska	_____
New York City, New York	_____	Death Valley, California	_____
Houston, Texas	_____	Mexico City, Mexico	_____
San Juan, Puerto Rico	_____	Atlanta, Georgia	_____

Detective Search: (There may be more than one answer to some of the questions.)

1. Which place on the map had the most rain or snow? _____

2. Which place on the map had the least amount of rain or snow? _____

3. Arctic Bay has about 10 days less snow or rain than _____.

4. On the northeastern coast of the North America which city had the most rain or snow? _____

5. Atlanta has about 5 more days of rain and snow than which city that borders one of the Great Lakes? _____

6. Which location in Canada gets the least amount of rain or snow? _____

7. San Francisco has about 6 times more rain or snow than what city in the western part of the United States? _____

8. Mexico City has about ½ as much rain or snow as _____.

9. Chicago has about 20 less days of rain or snow as _____.

©2004 by Incentive Publications, Inc., Nashville, TN.

Connecting literature with math skills

Teacher's Page

The Adventures of Flat Stanley

Enjoy this exciting activity where your Students can design their own "Flat Stanleys" and send them in the mail to their friends or relatives. It's an excellent way to integrate language arts, mathematics, and art.

Materials: *Flat Stanley* by Jeff Brown, large United States map, yarn, push pins, small 8" x 11½" student map, and an outline of Flat Stanley. Students can decorate their own Flat Stanleys according to where they plan to send him.

(1) Read *Flat Stanley* to your class and discuss with your children where they might like to send their Flat Stanleys. Give students the large outline of Flat Stanley and let them decorate him (markers, crayons, fabric, yarn, buttons, etc.) He will later be put in an envelope with a letter.

(2) Review letter writing skills. Students will write a letter to a friend or relative in other cities, states, or countries asking them to take Flat Stanley somewhere interesting in their community. Students will ask the receiver to take a photograph of Flat Stanley and return it with a brief summary of his travels.

(3) As soon as the letters are received, students can bring them with their photographs to share with their classmates. A special Flat Stanley wall or bulletin board can become the display area for the returned letters and photographs. Your students can mark on the large classroom United States map as well as their individual maps where Flat Stanley has traveled. Students can also measure distances and record them on their maps.

©2004 by Incentive Publications, Inc., Nashville, TN.

My Flat Stanley
by

_____

©2004 by Incentive Publications, Inc., Nashville, TN.

Fractions and Probability Skill Builders

NAME _____

Where in the World Can You Find Fractions?

Making a correlation between math and the "real"

Background Information: By relating math concepts to the real world, students can have a "hook" by which to internalize math concepts. Also, students have a reason for learning about numbers, fractions, measurement, geometry, etc. if they can see a connection to real-life situations or objects.

When introducing concepts such as fractions ask students to think of the many, varied, and unusual places fractions can be found in the "real" world. Make a classroom chart and leave it up while studying fractions.

A sample list of responses from a 2nd grade class:

1. Recipes
2. Shoe sizes
3. Road signs
4. Food package labels
5. Menus
6. Rulers
7. Bakeries
8. Sizes of clothes
9. Medicine bottles
10. Pencil sizes - $2\frac{1}{2}$
11. Computer games
12. Tire sizes
13. Weather forecasting
14. Tools
15. Airport signs
16. Test papers
17. Swimming scores
18. Sports equipment
19. Cartoons
20. Distances in space
21. Glasses' sizes
22. Measuring rain water
23. Discounts in sales
24. Billboards
25. Chef shows on TV
26. Ring sizes- $6\frac{1}{2}$
27. Measuring utensils

This activity took approximately 20 minutes. Each student then illustrated 4 of his/her favorites. This introductory lesson is also very effective when introducing whole numbers, standard measurement, geometry, money, and time.

©2004 by Incentive Publications, Inc., Nashville, TN.

Thinking of ways fractions are used in the "real world"

NAME _____

Fractions in the "Real World"

Think of many, varied, and unusual ways fractions are used in the "real world". Try to think of at least 10. The first one has been done for you.

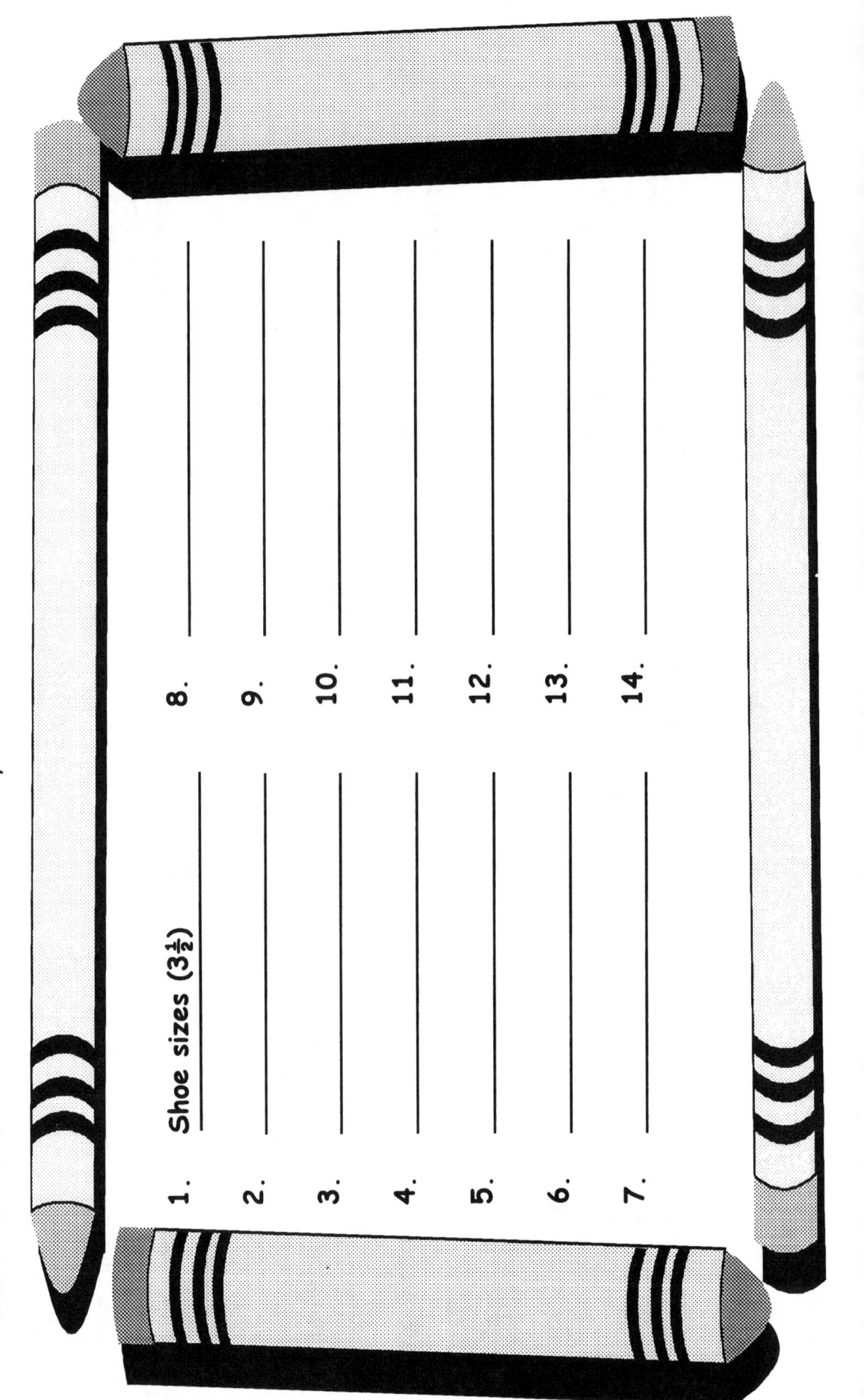

1. Shoe sizes (3½)
2. _____
3. _____
4. _____
5. _____
6. _____
7. _____
8. _____
9. _____
10. _____
11. _____
12. _____
13. _____
14. _____

63

©2004 by Incentive Publications, Inc., Nashville, TN.

NAME _____

When you can't fall asleep, why is it best to think about golf?

Write a fraction for the shaded part. Find your answer in the secret code. Each time your answer appears, write the letter of the problem above it.

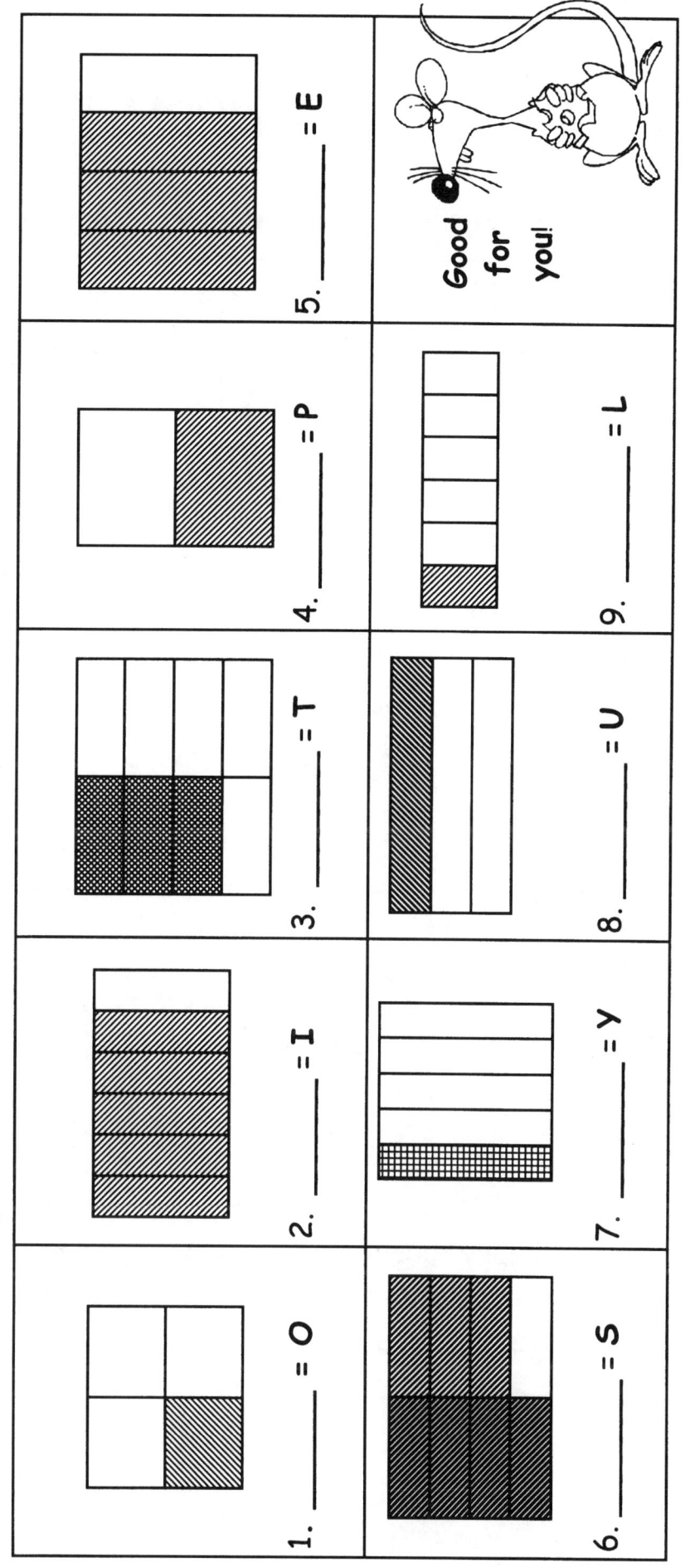

Naming fractions

1. ____ = O
2. ____ = I
3. ____ = T
4. ____ = P
5. ____ = E
6. ____ = S
7. ____ = Y
8. ____ = U
9. ____ = L

Good for you!

$\frac{5}{6}$ $\frac{3}{8}$ $\frac{1}{2}$ $\frac{1}{3}$ $\frac{3}{8}$ $\frac{7}{8}$ $\frac{1}{5}$ $\frac{1}{4}$ $\frac{1}{3}$ $\frac{3}{8}$ $\frac{1}{4}$ $\frac{7}{8}$ $\frac{1}{6}$ $\frac{3}{4}$ $\frac{1}{2}$

©2004 by Incentive Publications, Inc., Nashville, TN.

64

65 Choosing equal parts

NAME_____

PUT ON A HAPPY FACE!

Draw a happy face in the boxes that have figures with equal parts.

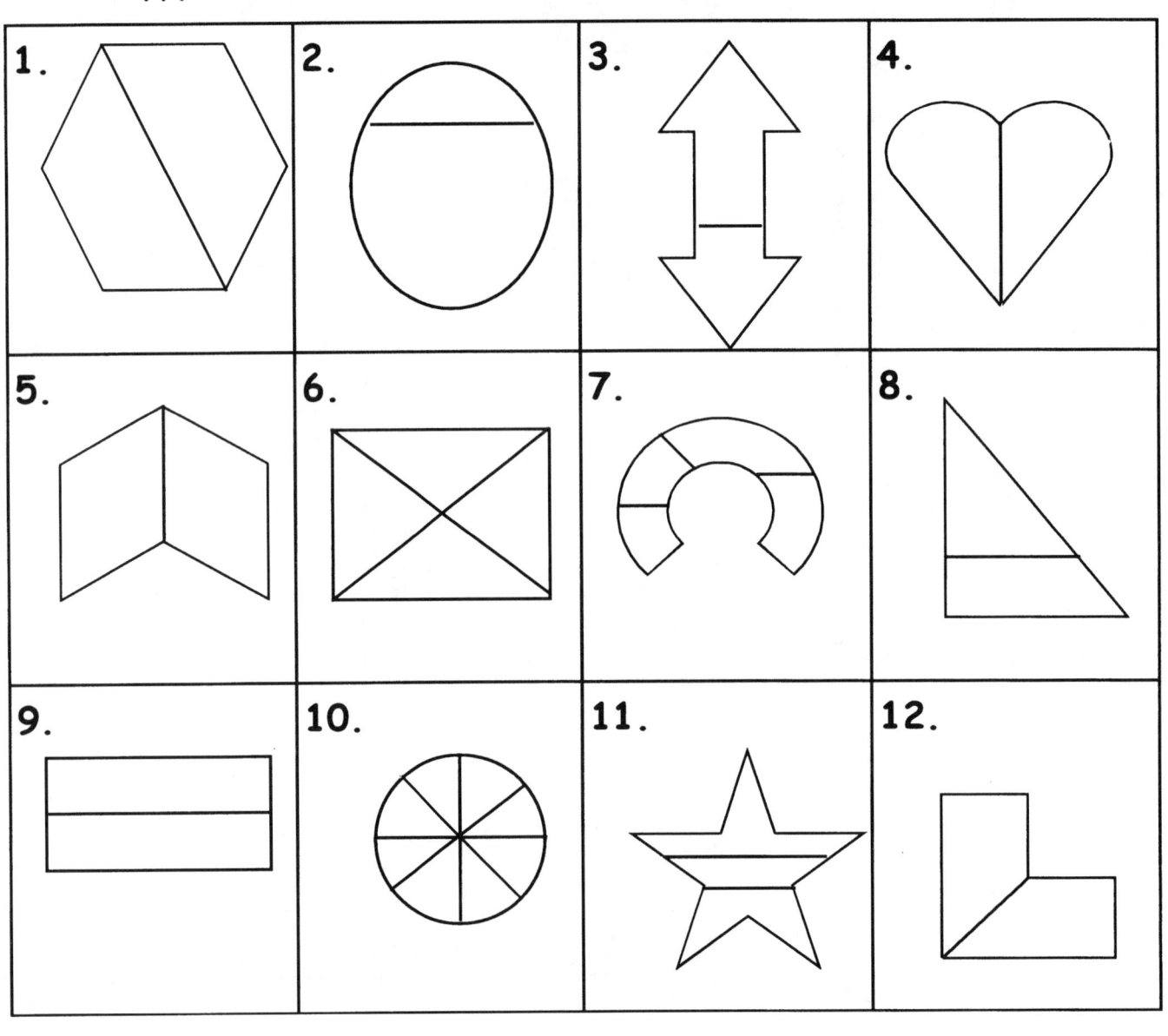

1.	2.	3.	4.
5.	6.	7.	8.
9.	10.	11.	12.

Did you draw 7 happy faces? Yes ___ No ___ If not, go back and check over your work. Now, draw 3 different figures and all three should have equal parts.

©2004 by Incentive Publications, Inc., Nashville, TN.

NAME _____

CLOWNIN' AROUND WITH FOURTHS!

The figure below is divided into fourths. Each part is called one-fourth. Together they made a whole.

| $\frac{1}{4}$ | $\frac{1}{4}$ | $\frac{1}{4}$ | $\frac{1}{4}$ |

Read each sentence and color parts of each figure.

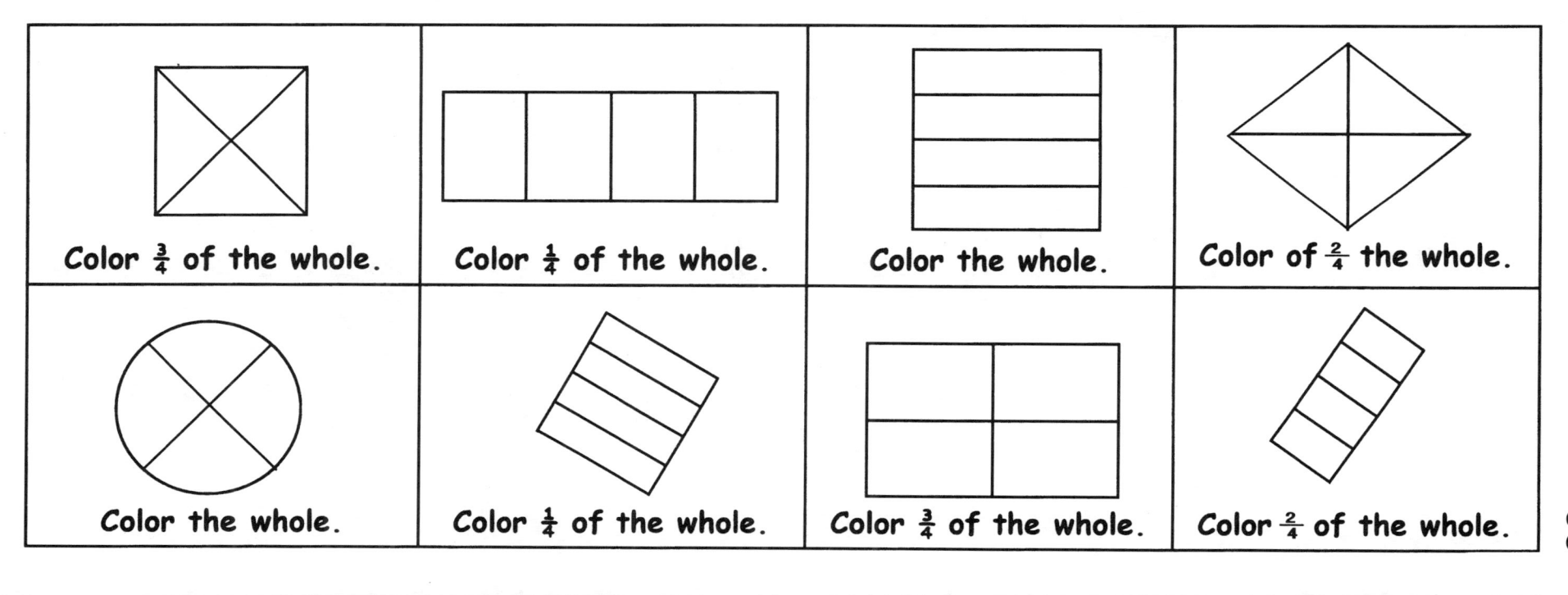

Color the whole.

Color $\frac{2}{4}$ of the whole.

Color the whole.

Color $\frac{2}{4}$ of the whole.

Color $\frac{1}{4}$ of the whole.

Color $\frac{3}{4}$ of the whole.

Color $\frac{1}{4}$ of the whole.

Color $\frac{3}{4}$ of the whole.

Color the whole.

Identifying fourths

©2004 by Incentive Publications, Inc., Nashville, TN.

67 Naming fractional parts   NAME_____

Let's Take a Ride!

1. How many total cars are on this octopus ride? _____
2. Color $\frac{1}{8}$ of the cars yellow.
3. Color $\frac{3}{8}$ of the cars red.
4. Color $\frac{2}{8}$ of the cars green.
5. Color $\frac{1}{8}$ of the cars purple
6. Color $\frac{1}{8}$ of the cars orange.

7. Circle the fraction that show the fewest number of cars.

    $\frac{1}{8}$ $\frac{3}{8}$

8. In which color car would you choose to ride?
_____

©2004 by Incentive Publications, Inc., Nashville, TN.

Identifying fractional parts of a set   NAME_____   68

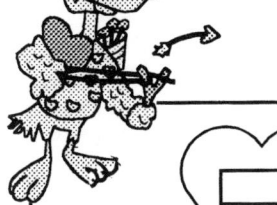

HEARTS GALORE!!

1.

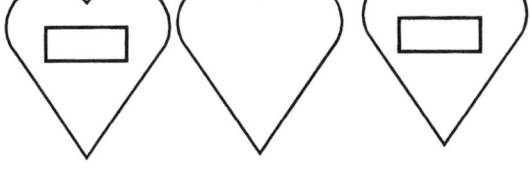

A. How many hearts in all? _____
B. How many have ▭? _____
C. What is the fraction? _____

2.

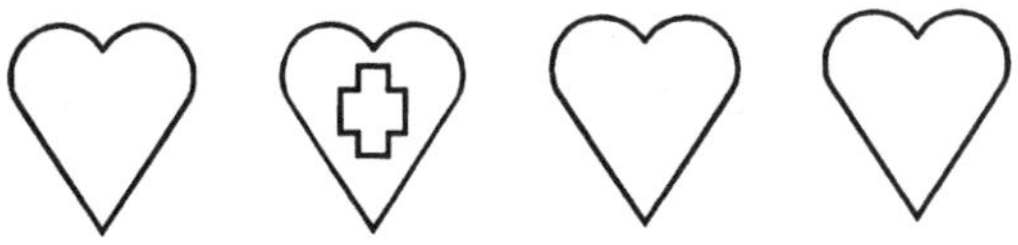

A. How many hearts in all? _____
B. How many have ✚? _____ 
C. What is the fraction? _____

3.

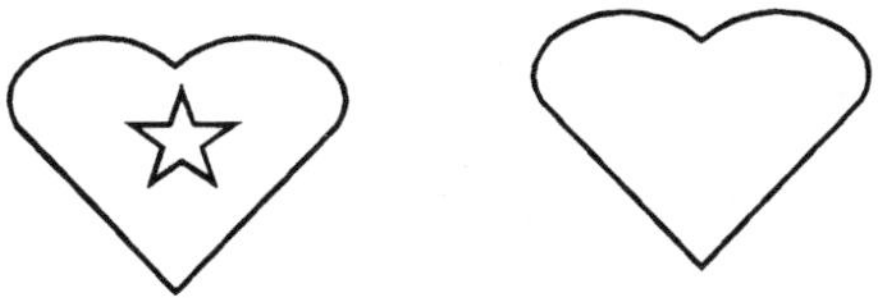

A. How many hearts in all? _____
B. How many have a ☆? _____ 
C. What is the fraction? _____

4.

A. How many hearts in all? _____
B. How many have a ⇕? _____ 
C. What is the fraction? _____

5.

A. How many hearts in all? _____
B. How many have a ↱? _____ 
C. What is the fraction? _____

6.

A. How many hearts in all? _____
B. How many hearts have a ◻? _____
C. What is the fraction? _____

7.

A. How many hearts in all? _____
B. How many have a ◯? _____ 
C. What is the fraction? _____

8.

A. How many hearts in all? _____
B. How many do not have a ◓? _____
C. What is the fraction? _____

Bonus: On the back draw 4 hearts and put a happy face in some of them. Write the fraction.

©2004 by Incentive Publications, Inc., Nashville, TN.

Identifying parts of a whole

CRAZY QUILTS

NAME _____

Write the fraction for each crazy quilt pattern. The first one has been done for you.

1. $\frac{3}{4}$

2. _____

3. _____

4. _____

5. _____

6. _____

7. _____

8. _____

Bonus: On the back or on another sheet of paper, design your own quilt and ask a friend to write in the fraction.

©2004 by Incentive Publications, Inc., Nashville, TN.

Finding fractional parts of a whole

NAME _____

70

What did the dolphin say when it bumped into the shark?

Solve each problem. Find your answer in the secret code. Each time you find your answer, write the letter of the problem above it.

1. $\frac{1}{4}$ of 28 = _____ (N)
2. $\frac{1}{8}$ of 16 = _____ (E)
3. $\frac{1}{3}$ of 18 = _____ (I)
4. $\frac{1}{7}$ of 21 = _____ (D)
5. $\frac{1}{6}$ of 6 = _____ (O)

6. $\frac{1}{5}$ of 25 = _____ (S)
7. $\frac{1}{3}$ of 30 = _____ (R)
8. $\frac{1}{2}$ of 22 = _____ (T)
9. $\frac{1}{9}$ of 36 = _____ (P)

__ __ __ __ __ __ , __ __ __ __
6 3 6 3 7 11 3 1 6 11

__ __ __ __ __ __ __ __ __ __
1 7 4 1 10 4 1 6 5 2

©2004 by Incentive Publications, Inc., Nashville, TN.

71 Using probability

NAME_____

MAYBE, MAYBE NOT!

The **chance** that something may or may not happen is called **probability**. What is the probability of each of the events happening below? Write the letter of your answer choice. The first one has been done for you.

| A. NEVER | B. PROBABLY NOT | C. MAYBE YES MAYBE NO | D. PROBABLY | E. ALWAYS |

1. You will have 2 hearts. _____A_____

2. There will be scattered showers tomorrow. _____

3. You will visit Mars by the time you are twelve. _____

4. You will have at least three meals today. _____

5. A dinosaur will visit your school tomorrow. _____

6. When you flip a coin, it will come up "heads". _____

7. A rainbow will appear in the sky today. _____

8. You will visit Antarctica next week. _____

9. The Atlanta Braves will win the World Series next year. _____

10. If you roll a die, you will get a "six". _____

On the back of this paper, write three statements like these. Ask a friend to fill in the blanks.

©2004 by Incentive Publications, Inc., Nashville, TN.

Determining probability

HEADS OR TAILS?

Toss a penny in the air. When it lands, look to see if it landed on heads or tails. Make a mark in the correct column after each toss. Toss the penny 20 times.

Your best guess: Out of 20 times, write how many times you think the penny will land heads and how many times tails.

Heads _____ **Tails** _____

Heads

Tails

	HEADS	TAILS		HEADS	TAILS
1st Toss			11th Toss		
2nd Toss			12th Toss		
3rd Toss			13th Toss		
4th Toss			14th Toss		
5th Toss			15th Toss		
6th Toss			16th Toss		
7th Toss			17th Toss		
8th Toss			18th Toss		
9th Toss			19th Toss		
10th Toss			20th Toss		

Using probability

What's in the Piñata?

73

NAME _____

Probability is the <u>chance</u> that something will or will not happen.

Color the items in the piñata. Use the color code.

Code for Color
balloons- 1 pink, 1 yellow, 1 purple
kazoos- 1 red, 1 blue
tambourine- brown
bracelet- purple
lollipops- 1 green, 1 red, 1 yellow
package of gum- orange

Give the probability that you will choose one of the items from the piñata. The first one has been done for you.

1. brown object $\underline{1}$ out of $\underline{11}$

2. kazoo _____ out of _____

3. purple object _____ out of _____

4. lollipop _____ out of _____

5. bracelet _____ out of _____

6. yellow object _____ out of _____

7. red object _____ out of _____

8. balloon _____ out of _____

9. package of gum _____ out of _____

©2004 by Incentive Publications, Inc., Nashville, TN.

Logical Reasoning Thinking Activities and Brain Teasers

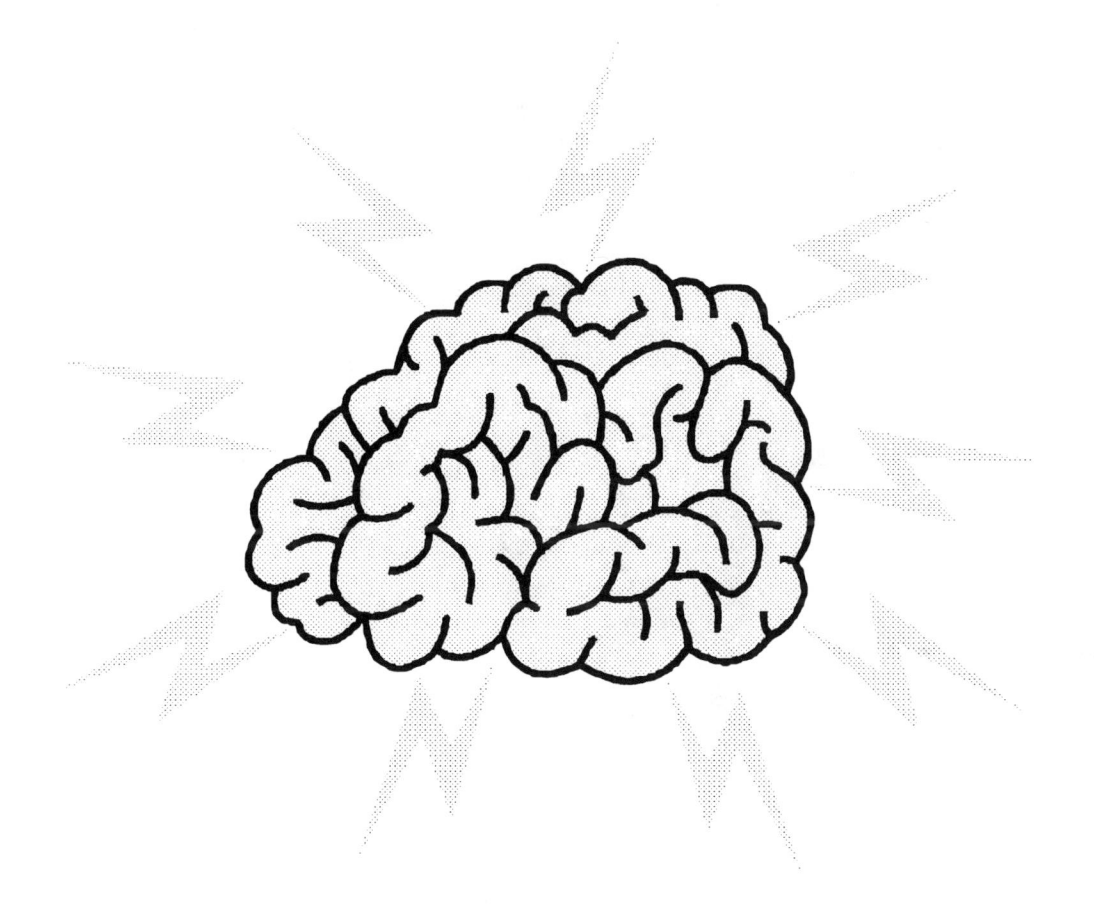

©2004 by Incentive Publications, Inc., Nashville, TN.

Using a variety of strategies to solve word problems

NAME _____

TRICKS FOR SOLVING WORD PROBLEMS!

1. UNDERSTANDING THE PROBLEM

It's important to:
- ☐ Read the problem carefully.
- ☐ Look for the important information.
- ☐ Jot down that information in your own words.

2. SOLVING THE PROBLEM

Think of strategies to try such as one or more of these:
- ☐ Use manipulatives.
- ☐ Act it out.
- ☐ Make a list or table.
- ☐ Look for a pattern.
- ☐ Use calculators or computers.
- ☐ Draw a diagram or picture.
- ☐ Make the problem easier by using smaller numbers.
- ☐ Close your eyes and visualize your word problem.
- ☐ Guess and check.
- ☐ Tell and print a story.
- ☐ Work backwards.

3. CHECKING THE PROBLEM

- ☐ Did you use all of the important information?
- ☐ Did you check your math carefully?
- ☐ Does your answer make sense?

Solving problems involving measurement   NAME_____   76

ANIMAL STATISTICS

Penguin
75 lb   36 in

Beaver
56 lb   30 in

Fox
14 lb   20 in

Woodpecker
2 oz   8 in

Panda Bear
300 lb   60 in

Koala
20 lb   24 in

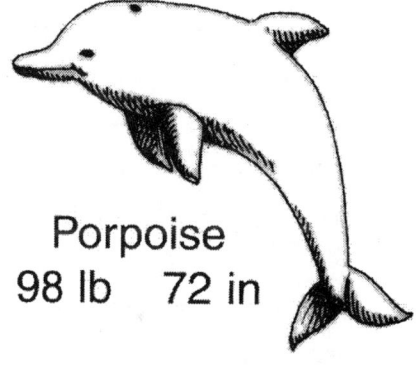

Spider
3 oz   4 in

Porpoise
98 lb   72 in

Trout
8 oz   9 in

Human baby
12 lb   29 in

Crab
10 oz   6 in

©2004 by Incentive Publications, Inc., Nashville, TN.

77 Using Logic to solve problems

NAME_____

Where does a bunny rabbit go when her coat needs grooming?

Use the chart, *Animal Statistics,* to help you solve the riddles below. Each time you solve a problem, find your answer in the secret code and write the letter of the problem above it.

1. I am an animal that weighs more than a beaver but less than a porpoise.
   What am I? _____ = (R)

2. I live in water and weigh less than 1 pound. I breathe through gills.
   What am I? _____ = (A)

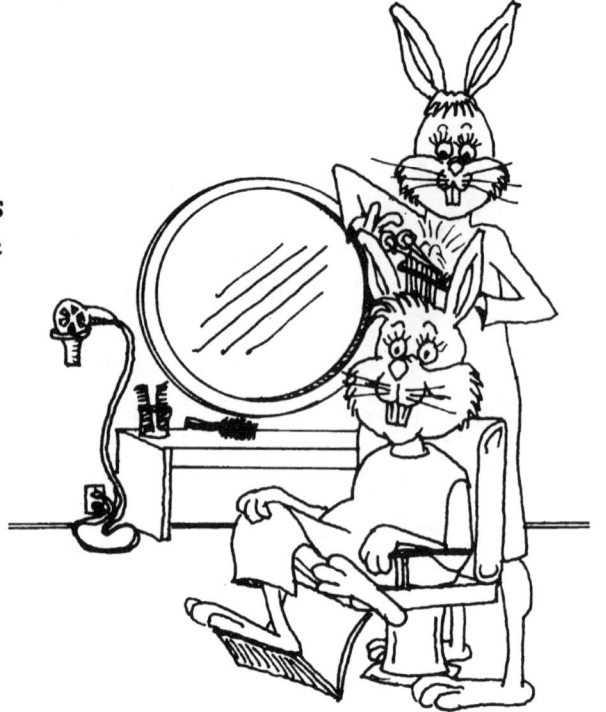

3. I am less than 5 inches long. I have eight legs and was a star in *Charlotte's Web*. What am I? _____ = (E)

4. I eat insects from trees with my sharp, pointed beak. I am very light in weight considering my size. What am I? _____ = (O)

5. I live in the ocean and am approximately $\frac{1}{2}$ foot long. I have an amazing ten legs. Many people think I'm pretty delicious to eat.
   What am I? _____ = (S)

6. I weigh less than a panda but more than a penguin. Many people think I am a fish but actually I am a mammal because I can be seen nursing my young.
   What am I? _____ = (D)

7. I have the most advanced brain of any animal on the chart. I have 2 eyes, 2 legs, and 2 arms. What am I? _____ = (T)

8. I am the heaviest and longest animal on the chart. I love bamboo and most people think I'm very cute. What am I? _____ = (H)

___	___	___	___	___	___	___
human baby	woodpecker	trout	panda	trout	penguin	spider

___	___	___	___	___	___	___
porpoise	penguin	spider	crab	crab	spider	penguin

©2004 by Incentive Publications, Inc., Nashville, TN.

Using logic

NAME_____

What's the Question?

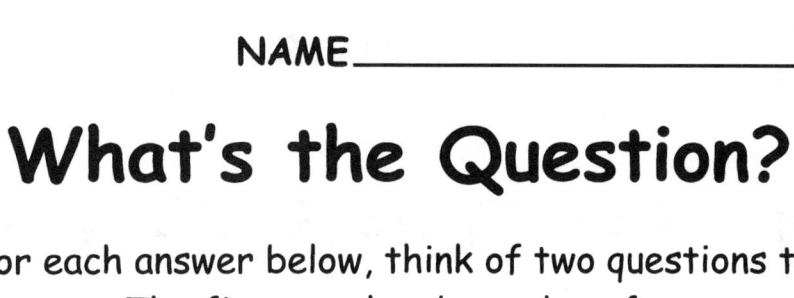

For each answer below, think of two questions that fit the answer. The first one has been done for you.

1. **The answer is 12. What is the question?**
   How many eggs in a dozen? How many months are in a a year?

2. **The answer is 10. What is the question?**
   _____
   _____

3. **The answer is 25. What is the question?**
   _____
   _____

4. **The answer is 8. What is the question?**
   _____
   _____

5. **The answer is 10¢. What is the question?**
   _____
   _____

6. **The answer is $\frac{1}{2}$. What is the question?**
   _____
   _____

7. **The answer is 2 feet. What is the question?**
   _____
   _____

©2004 by Incentive Publications, Inc., Nashville, TN.

What's the Question? - No. 2

For each answer below, think of two math questions that fit the answer. The first one has been done for you.

1. The answer is 30. What is the question? <u>How many days in June?</u> <u>How many minutes in ½ hour?</u>

2. The answer is 5. What is the question?
   _____
   _____

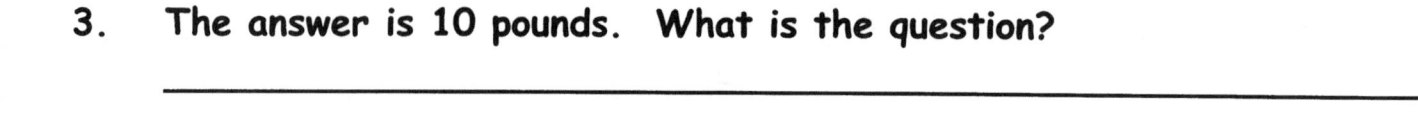

3. The answer is 10 pounds. What is the question?
   _____
   _____

4. The answer is 12:00 P.M. What is the question?
   _____
   _____

5. The answer is ¼. What is the question?
   _____
   _____

6. The answer is $1.00. What is the question?
   _____
   _____

7. The answer is 7. What is the question?
   _____
   _____

Using a calendar to solve problems

NAME _____

80

Birthdays of Famous Baseball Players

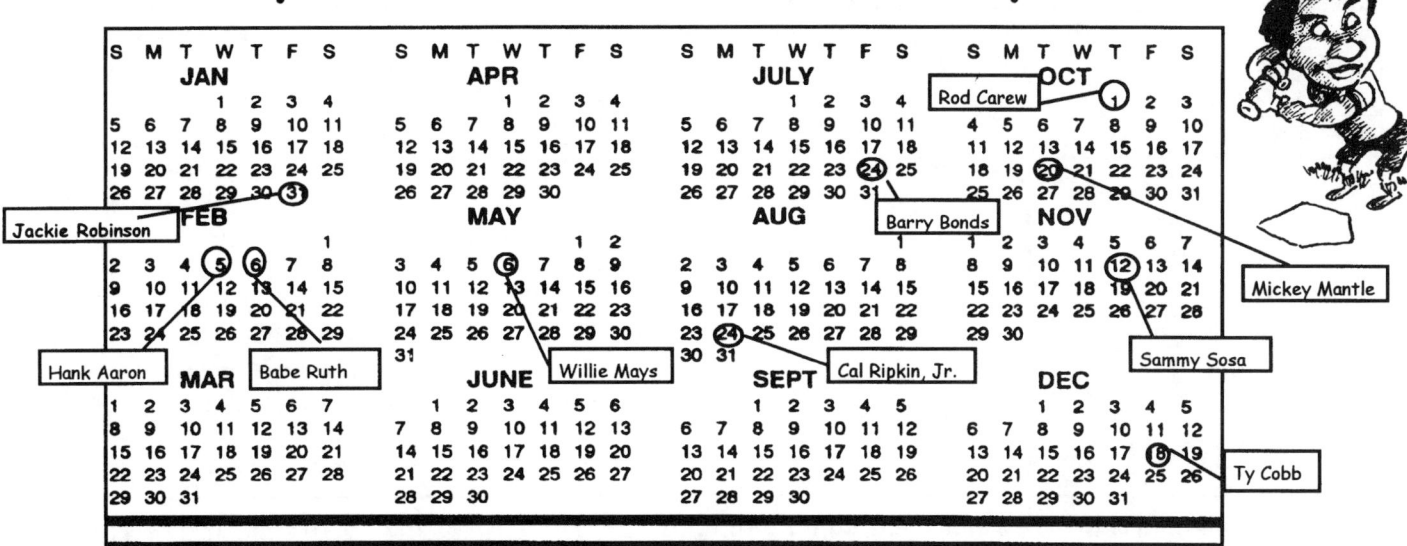

This calendar shows the birthdays of some famous baseball players. Use the calendar to help you answer the questions.

1. Which baseball player was born in the first month of the year? _____

2. Which two players have birthdays in the same week? _____

3. Which player was born on November 12? _____

4. Which player was born closest to July 4, our nation's birthday? _____

5. Which player was born in the last month of the year? _____

6. Which player was born on the 1st day of a month? _____

7. Which player was born in the 8th month? _____

8. Which player was born in the month after April? _____

9. Which player was born in October and has the same initials for his first and last name? _____

10. Who is your favorite baseball player? _____ Is his name on the calendar? Yes ____ No ____ Use the Internet to find out his birthday.

©2004 by Incentive Publications, Inc., Nashville, TN.

81 Using logical thinking when reading a calendar

NAME_____

What kind of insect do you swallow to relieve a cold?

Special Days
- New Year's Day – January 1
- Lincoln's Birthday – February 12
- Valentine's Day – February 14
- Washington's Birthday – Feb. 22
- St. Patrick's Day – March 17
- April Fool's Day – April 1
- Flag Day – June 14
- Independence Day – July 4
- Columbus Day – October 12
- Halloween – October 31
- Veteran's Day – November 11

Use the calendar above to help you solve the riddles. Each time you solve one of the riddles, find your answer in the secret code. Write the letter of the problem above the answer.

1. I am month that comes after October. Veteran's Day is in my month. What month am I?
   _____ = O

2. I am the month between the 1st and 5th months. I have 30 days and the 1st day is when you may be tricked by a friend. What month am I? _____ = E

3. I am a month with a few letters in my name. On the 14th day of my month people all over the United States fly the American flag with great pride. What month am I? _____ = A

4. I am a month that starts with a consonant and ends in "er". I am between the 7th and 10th months. What month am I? _____ = N

5. I am a month that has New Year's Day in me. What month am I? _____ = C

6. I am the between the 6th and 9th months of the year. My month has more than 4 letters but less 9 letters. What month am I? _____ = S

7. I am the month that has 2 presidents' birthdays. I have the fewest days in my month. What month am I? _____ = G

8. I am a month that comes after July but before December. I am the month where kids say, "Treat or Trick". What month am I? _____ = T

9. I am a spring month that has 31 days. On the 17th day most people wear green or they might be in danger of getting pinched. What month am I? _____ = D

___ ___ ___ ___ ___ ___ ___ ___ ___ ___ ___ ___
March April January November September February April August October June September October

©2004 by Incentive Publications, Inc., Nashville, TN.

Using logic in working with money

NAME_____

Case 1 – You're the Detective: What kind of potatoes are high in fat content?

DIRECTIONS: Find the answer to each set of clues. Find your answer in the secret code and write the letter of the problem above it. You will then have solved the case.

1. I am an amount that is an odd number. I am less than $.50. I am greater than $.20 + $.21. If you add my digits together they will total 7. What amount am I? _____ = A

2. I am an amount greater than 3 dimes but less than 2 quarters. I am an even number. I am an amount less than $.60 – $.15. Both of my digits are the same. What amount am I? _____ = O

3. I am two pennies less than 3 quarters. What amount am I? _____ = T

4. I am an amount between $.30 and $.50. I am more than 4 dimes. If you add my digits together, they will total 13. What amount am I? _____ = U

5. I am less than 3 dimes, 2 nickels, and 4 pennies. I am more than 1 dime, 5 nickels, and 3 pennies. If you add my digits together, they will total 6. What amount am I? _____ = H

6. I am an amount between $.60 and $.80. I am more than 3 quarters. My digits are the same. What amount am I? _____ = S

7. I am an amount that is less than $.80 – $.45. I am more than 1 quarter, 1 nickel and 2 pennies. If you add my digits, together they will be an even number. What amount am I? _____ = P

8. I am an amount that is more than 2 quarters, and 4 dimes. I am less than $1.00. Both of my digits are the same. What amount am I? _____ = C

9. I am more than $.50 and less than $.60. Both of my digits are the same. What amount am I? _____ = E

___ ___ ___ ___ ___   ___ ___ ___ ___ ___ ___ ___
$.99 $.44 $.49 $.99 $.42 $.33 $.44 $.73 $.43 $.73 $.44 $.55 $.77

©2004 by Incentive Publications, Inc., Nashville, TN.

83 Distinguishing time zones - logical thinking

NAME_____

IT'S ALL IN WHERE YOU LIVE!

PACIFIC TIME — 3:00 P.M.
MOUNTAIN TIME — 4:00 P.M.
CENTRAL TIME — 5:00 P.M.
EASTERN TIME — 6:00 P.M.

Directions:
Use the time zone map to help you solve the riddles. No answer is used more than once.

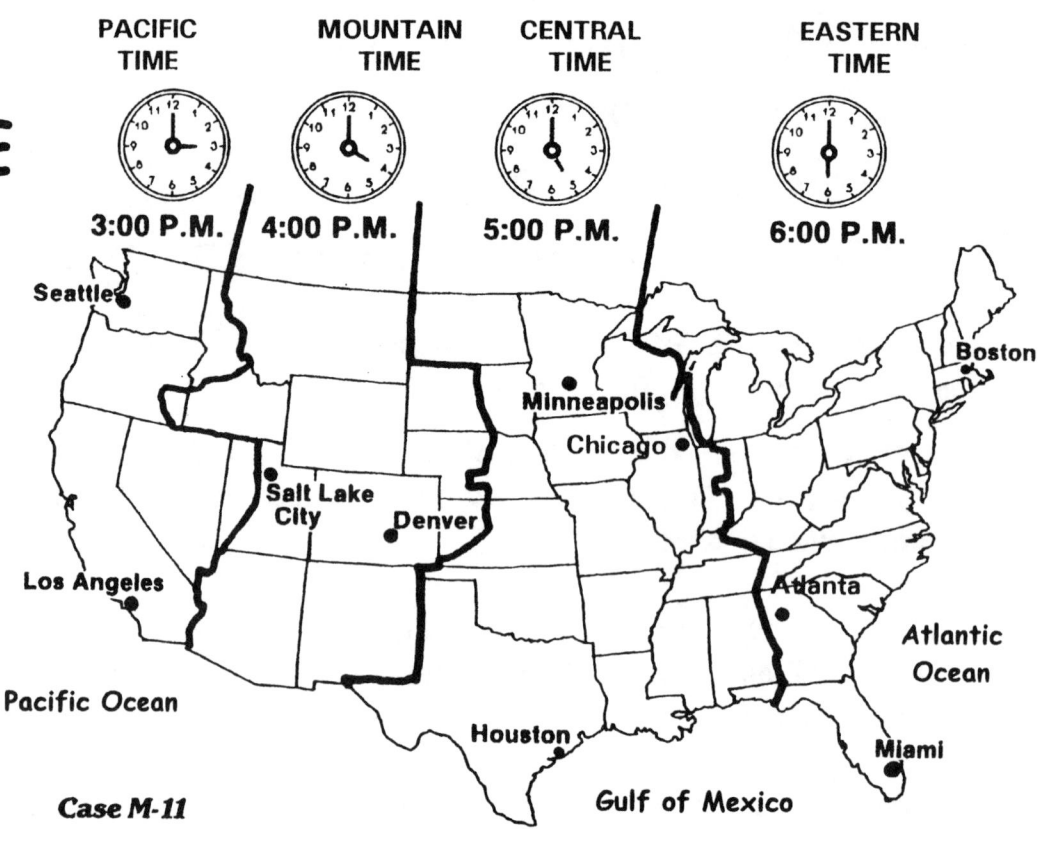

Case M-11

1. I am in the Eastern Time Zone. Part of my state is in the Central Time Zone but the largest part in the Eastern Time Zone. I am the city that is farthest south on the map. What city am I? _____

2. When it is 6:00 PM in Minneapolis, it is 5:00 PM in my zone. I am a city in Utah in the western edge of this time zone? What city am I? _____

3. When it is 4:00 PM in the Los Angeles, it is 6:00 PM in my time zone. I am a city located near the Great Lakes region. What city am I? _____

4. When it is 7:00 AM in Salt Lake City, it is 9:00 AM in my city. I am in the northeastern region of the United States. What city am I? _____

5. I am in the Central Time Zone and am located very near the Gulf of Mexico. What city am I? _____

6. What time zone do you live in? _____ Have you ever traveled to a different time zone? If so, write the name of the city/cities you were in and what time zone/zones they are in. *Write your answers on the back of this sheet.*

©2004 by Incentive Publications, Inc., Nashville, TN.

Using logic to solve number riddles

NAME_____

84

What noise does a nut make when it sneezes?

Solve each problem. Find your answer in the secret code and write the letter of the problem above it.

1. The Wood family arrived at 7:00 AM for their Amtrak trip to Washington D.C. Four trains arrived before theirs. Train A arrived at 8:00 AM. Train B arrived ½ hour after Train C and 2 hours before Train D. Train E arrived 1 hour after the latest train. Train C arrived 1 hour before Train A. When did the Wood family's Train E arrive? _____ = (A)

2. Mrs. Fields's class visited a small petting zoo. In the zoo there were 5 goats. The zoo had 1 more rabbit than ponies. There was 1 less rabbit than pot-bellied pigs and 1 more rabbit than chinchillas. The zoo had 3 more rabbits than goats. How many total animals were in the petting zoo? = _____ (S)

3. Five families decided to go on vacation the same week in July. The Opies departed at 11:00 AM. The Hoppers left 2 hours earlier than the Whitneys and ½ hour later than the Clarkes. The Lintons left 1½ hours later than the Clarkes and 2 hours later than the Opies. At what time did the Whitneys leave? = _____ (H)

4. At a garage sale, Mitchell found five Matchbox cars he wanted to purchase. The purple car costs $1.51. The orange costs 79¢ less than the green car, 34¢ less than the black car and $1.29 more than the red car. The red car cost 40¢ less than the purple one. How much did Mitchell spend on the 5 cars? =_____ (E)

5. Sally and her friends went camping and while they were there they went fishing. Sally caught 8 fish and Laura caught ½ as many as Antonio. Sol caught 1 more than Antonio and 3 more than Carlos and 1 less than Sally. Carlos caught ½ as many as Sally. How many total fish did Sally and her friends catch? = _____ (W)

6. Mrs. Bixby's class was practicing their long jumps. Barry jumped 49 inches. Paul jumped 8 inches more than Amanda, 3 inches more than Leslie and 2 inches less than Matt. Leslie jumped 6 inches less than Barry. How far did Matt jump? = _____ (C)

_____ _____ _____ _____ _____ _____
  48   10:30   36   2:00 $10.95   28

©2004 by Incentive Publications, Inc., Nashville, TN.

85 Using logic to solve word problems

NAME_____

What do aliens use to tie up spacemen?

1. Mrs. Whitney enjoyed making jelly. She made a half-dozen jars of grape jelly and a dozen jars of blueberry jelly. She made twice as many jars of peach jelly as blueberry jelly and 3 less jars of apricot than blueberry. How many total jars did she make? _____ = T

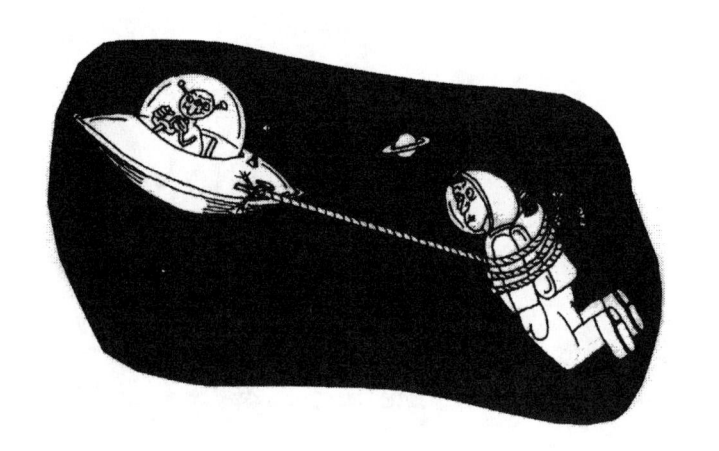

2. Drew and his friends decided to catch fireflies at their church picnic. Drew caught 8 fireflies. Meredith caught 5 more fireflies than Drew. Mitchell caught 13 more fireflies than Cate. Cate caught 3 times as many as Meredith. Kevin caught 10 more fireflies than Drew. What was the total number of fireflies caught by Drew and his friends? _____ = O

3. Six go carts lined up for the soapbox derby. The black car went 26 laps before making a pit stop. The striped car went 1 more lap than the yellow car, 3 less laps than the hot pink car, and 2 more laps than the black car. The orange car went 4 more laps than the black car and 1 less lap than the purple car. How many total laps did all the cars make before their pit stops? _____ = S

4. Grant had his own pet sitting service during the summer months. Mr. Sims paid him $8.75. Mr. Bailey paid him $1.00 more than Mr. Sims and 85 cents more than Mrs. Clements. Mrs. Bradley paid Grant $1.30 more than Mr. Bailey. How much total money did Grant make with his pet sitting service? _____ = N

5. The Drama Club held a bake sale. The cakes sold for $3.50 each. A bag of cookies sold for half less than the cakes and $.30 less than ½ dozen cupcakes. Brownies sold for $.25 each and 3 doughnuts sold for $1.00 less than the ½ dozen cupcakes. One customer bought a cake, a bag of cookies, 6 cupcakes, 2 brownies, and 3 doughnuts. How much did the Drama Club earn from this one customer? _____ = K

6. Several cities in the South were experiencing a drought. During the one week in July, Atlanta got 1 inch of rain, Nashville got 1 inch more rain than Charleston, 2 more inches than New Orleans, and 3 more inches than Atlanta. Savannah got 2 inches less than Nashville. How much total rainfall did all five cities receive? _____ = A

7. John and his friends ordered small individual pizzas. John's pizza had 4 toppings, Fred had one more topping than Tim and 4 less than Matt. Mike had 2 less toppings than John and 1 more than Tim. How many total toppings did the boys have on their pizzas? _____ = R

_____ _____ _____ _____ _____ _____ _____ _____ _____ _____
 12   173   51   15   130 $8.85 $38.45   130   51   173

©2004 by Incentive Publications, Inc., Nashville, TN.

Using logic to solve number riddles

NAME_____

86

What did they call Old MacDonald when he joined the army?

Solve each problem. Find your answer in the secret decoder and write the letter of the problem above it.

1. Laura finished 18 addition problems. Robert finished 3 more problems than Erin and 4 less than Hannah. Tah-Toh finished 2 more problems than Robert and 3 less than Laura. How many total math problems did all five students complete? _____ (O)

2. Nicole planned to make goody bags for her birthday party. She separated her 2 dozen items into piles. She had 5 balloons, 2 more kazoos than chocolate candies and 1 more bubblegum than fingernail decorations. Nicole had 1 more balloon than the chocolate candies. How many fingernail decorations did Nicole have? = _____ (E)

3. Tah-Tah and her family ordered waffle fries. The 1st box held 26 fries. The fourth box held 2 dozen waffle fries. The 3rd box held 5 more fries than the 5th box and 2 more fries than the 2nd box. The 4th box held 6 more fries than the 2nd box. How many did the 3rd box hold? = _____ (I)

4. Maddie, Tyler, Amber, Matt and David each had hamsters. Maddie's hamster had 5 babies. Tyler's hamster had 2 less babies than Matt's. Amber's hamster had 1 less baby than David's and 1 more than Tyler's. David's hamster had 6 babies. What was the total number of hamster babies born? = _____ (G)

5. Elizabeth liked to make sundaes. After making one for herself she decided to make 3 more for her friends. The ice cream cost $1.89. Strawberries cost 50¢ more than the bananas and 39¢ less than the hot fudge. The nuts cost 26¢ more than the strawberries and 60¢ less than the ice cream. How much did Elizabeth spend to make her sundaes? = _____ (J)

____ ____ ____ ____ ____ ____ ____
 26   20   26   20 $6.16 73   4

©2004 by Incentive Publications, Inc., Nashville, TN.

The Bake Sale!

Lily and Samantha wanted to bake 1 cake, 1 pie, and 1 dozen cookies for the bake sale. How much of each item do they need to make the three items?

Come to the Bake Sale!

Chocolate Chip Cake
(serves 6)

$1\frac{1}{2}$ cups of flour

2 eggs

$\frac{3}{4}$ cup sugar

$\frac{1}{4}$ cup of oil

2 cups of chocolate icing

Lemon Icebox Pie
(serves 8)

1 cup of condensed milk

3 tablespoons of lemon juice

3 eggs

$\frac{1}{2}$ cup sugar

$1\frac{3}{4}$ cup of graham crackers

Chocolate Chip Cookies
(serves 12)

$1\frac{1}{4}$ cup of flour
1 cup of water
$\frac{3}{4}$ cup of sugar
$\frac{1}{2}$ cup of milk
$\frac{1}{4}$ cup of oil
1 cup of chocolate chips
1 egg

1. Sugar _____ cups

2. Flour _____ cups

3. Chocolate icing _____ cups

4. Graham crackers ___ cups

5. Eggs _____

6. Oil _____ cups

7. Lemon juice _____ tablespoons

8. Chocolate chips _____ cups

BONUS: Circle the total number of cups used to make all three desserts:

11½   12½   13½

Using logic to solve problems NAME_____ 88

It's Show Time!!

You and your family are planning your trip to *Water World*. You want to see all four of the main shows. Each show is scheduled at least twice a day. You need to plan 45 minutes for lunch. Decide in which order you will see the shows and complete the *Water World* planning guide at the bottom of the page.

Freddy, the Dolphin
30 minute show

Kasey, the Killer Whale
45 minute show

Ollie, the Otter
30 minute show

Justine, the Seal
45 minute show

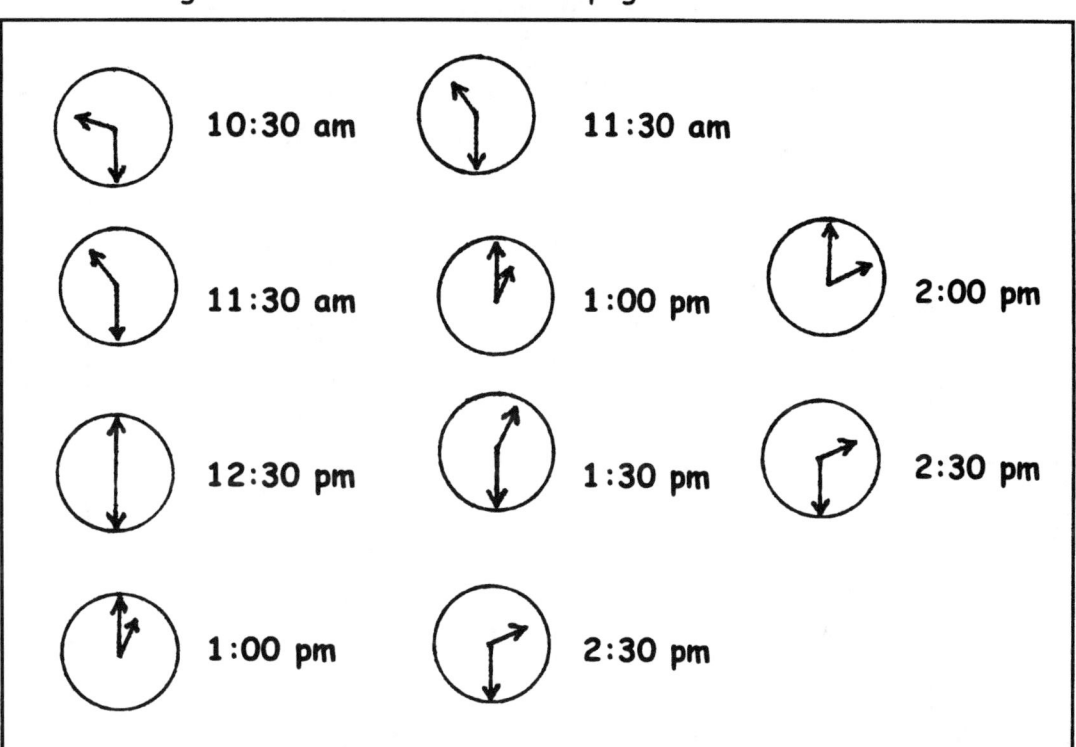

10:30 am	11:30 am	
11:30 am	1:00 pm	2:00 pm
12:30 pm	1:30 pm	2:30 pm
1:00 pm	2:30 pm	

Water World Planning Guide

SHOW	TIME
_____	_____
_____	_____
Lunch _____	_____
_____	_____
_____	_____

©2004 by Incentive Publications, Inc., Nashville, TN.

89 Logical Reasoning-Money

NAME _____

THE MONEY CHALLENGE!

Help Casey the Centipede figure out this money challenge. Read the clues to find what coins are used. Write the coins on each line and then add those coins together to find the total. Use Q for quarter, D for dime, N for nickel, and P for penny. Circle your answer at the bottom of the page.

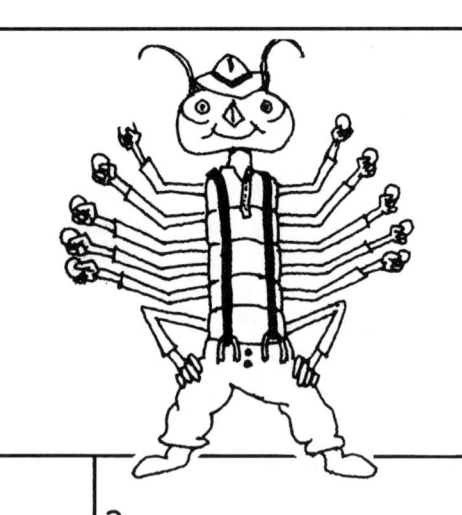

1. 1Q + 2N + 2P + 1D = 47¢ There are: • 6 coins • 4 different kinds • An equal number of pennies and nickels • 1 dime	2. _____ = ____ There are: • 6 coins • 2 more dimes than pennies • No nickels or quarters	3. _____ = ____ There are: • 6 coins • 3 different kinds • No dimes • 1 nickel more than quarters • 1 penny
4. _____ = ____ There are: • 7 coins • 2 nickels • More dimes than quarters • An equal number of pennies and nickels	5. _____ = ____ There are: • 5 coins • No dimes • 2 more nickels than quarters or pennies	6. _____ = ____ There are: • 7 coins • Same number of dimes, quarters, and nickels
7. _____ = ____ There are: • 7 coins • 3 quarters • 2 less dimes than quarters • Same number of pennies as quarters	8. _____ = ____ There are: • 9 coins • Four different kinds • 2 quarters • Same number of pennies as nickels • 1 dime	9. _____ = ____ There are: • 10 coins • 4 different kinds • 4 pennies • Same number of quarters, nickels, and dimes

84¢   42¢   78¢   (47¢)   88¢   57¢   66¢   41¢   81¢

©2004 by Incentive Publications, Inc., Nashville, TN.

ANSWER KEY

2. Answers will vary.
3. Products will vary.
4. Clocks will vary.
5. Schedules will vary.
6. 1. 7:30 AM 2. 8:30 AM 3. 9:30 AM 4. 11:00 AM 5. 12:00 AM 6. 1:30 PM 7. 3:30 PM
7. 1. 7:05 AM 2. 7:10 AM 3. 7:30 AM 4. 7:40 AM 5. 7:50 AM 6. 8:00 AM
   7. 8:10 AM 8. 8:25 AM 9. 8:45 AM 1. 105 minutes 2. Yes, she arrives by 8:45.
8. Schedules will vary.
9. 1. 5:30 AM 2. 6:00 AM 3. 6:15 AM 4. 8:15 AM 5. 10:30 AM 6. 12:45 PM 7. 3:00 PM
   8. 5:00 PM
10. Some possible answers: 1. 4:15, four fifteen, 15 minutes after 4 2. 5:45, five forty-five, 15 minutes before 6 3. 2:50, two fifty, 10 minutes before 3 4. 8:10, eight ten, 10 minutes after 8 5. 10:25, ten twenty-five, 25 minutes after 10 6. 6:40, six forty, 20 minutes until seven 7. 12:15, twelve fifteen, 15 minutes after 12 8. 1:35, one thirty-five, 25 minutes before 2 9. 9:20, nine twenty, 20 minutes after 9 10. 7:55, seven fifty-five, 5 minutes until 8 11. 3:30, three thirty, 30 minutes after 3, 30 minutes before 4 12. 11:45, eleven forty-five, 15 minutes till 12.
11-12. Students' MATHO boards will vary.
13. January, February, March, April, May, June, July, August, September, October, November, December
14-15. Time lines will vary.
16. 1. Blue Jeans, 1850 2. 123 years 3. CD Player, about 20 years ago; Coca-Cola, about 120 years ago; Barbie Doll, about 45 years ago; Air Bags, about 30 years ago; Airplane, about 100 years ago; Basket ball, about 110 years ago 4. CD Player 5. Answers will vary. 6. Answers will vary.
17. 1. 40¢, notebook paper 2. 22¢, ruler 3. 31¢, glue 4. 25¢, eraser 5. 36¢, crayons 6. 20¢, pencil 7. 33¢, scissors
18. 1. nickel 2. penny 3. dime 4. nickel 5. nickel 6. nickel 7. penny 8. nickel 9. nickel
19. 1. 1 quarter, 1 dime, 1 nickel, 1 penny 2. 2 dimes 3. 1 dime, 1 penny 4. 2 quarters 5. 1 dime, 1 nickel 6. 1 nickel
20. 1. Construction Paper, 9¢ 2. Pipe cleaners, 12¢ 3. Paint brushes, 86¢ 4. Watercolors, $2.68 5. Scissors, 22¢ 6. Beads, 3¢ Bonus: $18.42
21. Possible answers: 1q, 1d, 1n, 0p, 3 coins; 1q, 0d, 2n, 5p, 8 coins; 0q, 3d, 2n, 0p 5 coins; 0q, 3d, 1n, 5p 9 coins; 0q, 2d, 4n, 0p, 6 coins.
22. Possible answers– 1. 1q, 2n; 2. 4n; 3. 1q, 4p; 4. 3d; 5. 2q, 1p; 6. 2d, 1n, 1p; 7. 1q, 2d, 1n, 2p
23. 1. 50¢ 2. 37¢ 3. 40¢ 4. 31¢ 5. 45¢ 6. 48¢
24. Possible answers: 1. 0q, 1d, 1n, 3p 2. 1q, 0d, 1n, 3p 3. 2q, 0d, 0n, 4p 4. 0q, 1d, 0n, 4p 5. 3q, 0d, 0n, 3p 6. 0q, 4d, 4n, 0p 7. 0q, 1d, 2n, 2p 8. 1q, 1d, 0n, 3p 9. 0q, 1d 3n, 0p 10. 1q, 2d, 4n, 0p 11. 0q, 0d, 1n, 6p 12. 1q, 1d, 1n, 2p 13. 4q, 1d, 2n, 1p 14. 4q, 2d, 2n, 4p
25. 1. 45¢ 2. 65¢ 3. 41¢ 4. 35¢ 5. Vehicles will vary
26. 1. $5.00 2. $9.00 3. $13.00 4. $5.00 5. $4.00 6. $6.00 7. $1.00
    Total cost of train: $43.00 Bonus: Yes, The train set is $43.00 and the birthday money is $50.00
27. Answers will vary.
28. 1. Correct 2. Incorrect, $11.19 3. Correct 4. Correct 5. Correct 6. Incorrect, $2.12 7. Incorrect, $5.63 8. Correct 9. Incorrect, $1.93
29. 1. $2.55 2. $1.95 3. $.25 4. $3.00 5. $2.67 6. $1.70 7. $6.04 Answer: LITTLE BOO PEEP
30. 1. $.35 2. $.43 3. $1.59 4. $3.56 5. $.78 6. $.74 7. $4.57 8. $.76 9. $2.21 10. $2.01 11. $.88 12. $7.75 13. $4.01 Answer: ITS PETALS WERE BROKEN
31. 1. $.24 2. $.34 3. $.89 4. $.21 5. $2.11 6. $.78 7. $.11 8. $1.41 9. $.56 10. $.98 11. $.33 12. $.18 13. $.66 14. $2.67 THEY BECAME TONGUE TIED
32. 1. $1.23 2. $.78 3. $1.21 4. $3.90 5. $.86 6. $1.11 7. $.65 8. $6.71 9. $.48 10. $5.95 11. $2.17 12. $8.32 I DON'T HAVE A SCENT AND IT STINKS
33. Answers will vary.
34. Problems will vary.
37. 1. 2 2. 100 3. 48 4. lb 5. 4 6. 5 7. 3 8. 9 9. 1 10. 28 11. oz ON THEIR BUNNYMOON
38. 1. 2 2. 1 3. 8 4. 2 5. 1 6. 4 7. 8 8. 8
39. Charts will vary.
40. 1. 1 2. 2 3. 3 4. 4 5. 8 6. 6 7. 4,000 8. 32 IN A RIVER BANK
41. 1. 12 2. 3 3. 6 4. 1 5. 18 6. 5 7. 36 8. 2 9. 9 SO IT DOESN'T PEEL

©2004 by Incentive Publications, Inc., Nashville, TN.

91

42. 1. 2 2. 120 3. 52 4. 180 5. 200 6. 1 7. 50 8. 21 9. 4 10. 5 11. 10 12. 36 13. 730 14. 30
SHE DRIBBLES ALL OVER THE PLACE

43. 1. 13 cm 2. 12 cm 3. 8 cm 4. 9 cm 5. 6 cm 6. 10 cm 7. 4 cm 8. 2 cm 9. 15 cm

44. Drawings will vary.

45. 1. Gram 2. Meter 3. Liter 4. Meter 5. Meter/Centimeter 6. Gram 7. Meter 8. Liter 9. Gram 10. Liter 11. Meter 12. Gram

46. Answers will vary.

47. Measurement of Length: 1. Meter/Centimeter 2. Millimeter 3. Kilometer 4. Centimeter 5. Meter/Centimeter 6. Centimeter or Millimeter 7. Meter Liquid Measurement: 1. Milliliter 2. Liter 3. Liter 4. Milliliter 5. Liter 6. Milliliter 7. Liter Measurement of Weight: 1. Gram/Kilogram 2. Kilogram 3. Milligram 4. Gram/Kilogram 5. Milligram 6. Kilogram 7. Milligram

48. Answers will vary.

49-51. 1. 4 cm 2. 4 cm 3. 6 cm 4. 6 cm 5. 22 cm 6. 7 cm 7. 17 cm 8. 16 cm 9. 11 cm 10. 15 cm 11. 18 cm 12. 13 cm

52. 1. 20° 2. 0° 3. 37° 4. 68° 5. 32° 6. 98.6° 7. C 8. F 9. C 10. F 11. F 12. C

53-54. 1. Parasaurolophus 2. Triceratops 3. Stegosaurus 4. Velociraptor 5. Deinonychus 6. Pterosaurus 7. Tyrannosaurus 8. Iguanadon 9. Brontosaurus What Am I? TRICERATOPS

55-56. 1. Atlanta, Georgia, 72, 51, 21° 2. Minneapolis, Minnesota, 54, 35, 19° 3. San Francisco, California, 62, 50, 12° New York City, New York, 57, 42, 15° Dallas, Texas, 75, 55, 20° 2. Phoenix, Burlington, 55°

57-58. 1. Toronto, 40; New York City, 35; Houston, 25; San Juan, 50 2. Fairbanks, 20; Death Valley, 5; Mexico City, 15; Atlanta, 35 Detective Search: 1. Vancouver 2. Death Valley 3. Houston 4. Halifax 5. Chicago 6. Arctic Bay 7. Death Valley 8. San Francisco, Chicago, Winnipeg 9. San Juan

59-60. Student drawings and letters will vary.

63. Answers will vary.

64. 1. $\frac{1}{4}$ 2. $\frac{5}{6}$ 3. $\frac{3}{8}$ 4. $\frac{1}{2}$ 5. $\frac{3}{4}$ 6. $\frac{7}{8}$ 7. $\frac{1}{5}$ 8. $\frac{1}{3}$ 9. $\frac{1}{6}$ IT PUTTS YOU TO SLEEP

65. 1. ☺ 4. ☺ 5. ☺ 6. ☺ 9. ☺ 10. ☺ 12. ☺

66. Colorings will vary.

67. 1. 8 7. $\frac{1}{8}$ Colorings will vary.

68. 1. 3, 2, $\frac{2}{3}$ 2. 4, 1, $\frac{1}{4}$ 3. 2, 1, $\frac{1}{2}$ 4. 6, 5, $\frac{5}{6}$ 5. 4, 3, $\frac{3}{4}$ 6. 6, 2, $\frac{2}{6}$ 7. 3, 1, $\frac{1}{3}$ 8. 8, 6, $\frac{6}{8}$

69. 1. $\frac{3}{4}$ 2. $\frac{3}{8}$ 3. $\frac{1}{2}$ 4. $\frac{7}{10}$ 5. $\frac{2}{3}$ 6. $\frac{8}{12}$ 7. $\frac{4}{9}$ 8. $\frac{4}{5}$

70. 1. 7 2. 2 3. 6 4. 3 5. 1 6. 5 7. 10 8. 11 9. 4   I DIDN'T DO IT ON PORPOISE

71. Answers will vary.

72. Answers will vary.

73. 1. 1 out of 11 2. 2 out of 11 3. 2 out of 11 4. 3 out of 11 5. 1 out of 11 6. 2 out of 11 7. 2 out of 11 8. 3 out of 11 9. 1 out of 11

76-77. 1. Penguin 2. Trout 3. Spider 4. Woodpecker 5. Crab 6. Porpoise 7. Human baby 8. Panda TO A HARE DRESSER

78. Answers will vary.

79. Answers will vary.

80. 1. Jackie Robinson 2. Hank Aaron, Babe Ruth 3. Sammy Sosa 4. Barry Bonds 5. Ty Cobb 6. Rod Carew 7. Cal Ripkin, Jr. 8. Willie Mays 9. Mickey Mantle

81. 1. November 2. April 3. June 4. September 5. January 6. August 7. February 8. October 9. March DECONGEST-ANT

82. 1. $.43 2. $.44 3. $.73 4. $.49 5. $.42 6. $.77 7. $.33 8. $.99 9. $.55 COUCH POTATOES

83. 1. Miami 2. Salt Lake City 3. Minneapolis 4. Boston 5. Houston

84. 1. 10:30 2. 36 3. 2:00 4. $10.95 5. 28 6. 48 CASHEW

85. 1. 51 2. 130 3. 173 4. $38.45 5. $8.85 6. 12 7. 15 ASTROKNOTS

86. 1. 73 2. 4 3. 20 4. 26 5. $6.16 GI GI JOE

87. 1. 2 2. 2$\frac{1}{4}$ 3. 2 4. 1$\frac{1}{4}$ 5. 6 6. $\frac{1}{2}$ 7. 3 8. 1 BONUS: 12$\frac{1}{2}$ cups

88. Schedules will vary. Example of a workable schedule: 10:30 a.m.- 11:00 a.m. (Freddy the Dolphin), 11:30 a.m.-12:15 p.m. (Kasey, the Killer Whale), 12:15 p.m.-1:00 p.m. (Lunch), 1:00-1:45 (Justine, the Seal), 2:30 p.m. - 3:00p.m. (Justine the Seal)

89. 1. 1Q + 2N + 2P + 1D = 47¢ 2. 4D + 2P = 42¢ 3. 3N + 2Q + 1P = 66¢ 4. 2N + 2P + 2D + 1Q = 57¢ 5. 3N + 1Q + 1P = 41¢ 6. 2D + 2Q + 2N + 1P = 81¢ 7. 3Q + 1D + 3P = $.88 8. 2Q + 1D + 3P +3N = 78¢ 9. 4P + 2Q + 2D + 2N = 84¢

©2004 by Incentive Publications, Inc., Nashville, TN.